D1743805

PEOPLE AND OTHER ANIMALS:
REAL DIFFERENCES, REAL SIMILARITIES

PEOPLE AND OTHER ANIMALS:
REAL DIFFERENCES, REAL SIMILARITIES

Norma A. Donner

with a Foreword by

Marty Stouffer

BARRICADE
BOOKS
Fort Lee, New Jersey

Published by Barricade Books Inc.
185 Bridge Plaza North
Suite 308-A
Fort Lee, NJ 07024
www.barricadebooks.com

Copyright © 2002 by Norma A. Donner
All Rights Reserved.

No part of this book may be reproduced, stored in a retrieval system, or transmitted in any form, by any means, including mechanical, electronic, photocopying, recording, or otherwise, without the prior written permission of the publisher, except by a reviewer who wishes to quote brief passages in connection with a review written for inclusion in a magazine, newspaper, or broadcast.

Library of Congress Cataloging-in-Publication Data

Donner, Norma A.
    People and other animals : real differences, real similarities / Norma A. Donner ; with an introduction by Marty Stouffer.
        p. cm.
    Includes bibliographical references.
    ISBN 1-56980-236-X (casebound)
        I. Animal behavior. 2. Human ecology. 3. Physiology, Comparitive. I. Title.

QL751 .D6155 2002
591.5--dc21
                                                                2002026062

First Printing
Manufactured in the United States of America

TABLE OF CONTENTS

Dedication

In loving memory of *JOAN MOFFAT WILSON*
(7/10/1922—12/16/96)

———————

"Neither Beasts Without nor Beasts Within
are as beastly as they have been painted."

—Mary Midgley
Beast and Man

FOREWORD
BY MARTY STOUFFER

I've always considered myself a storyteller. I started making films because I wanted others to share my enthusiasm for animals. In this way, I've done a great deal to encourage others to share my interests. Yet until I read this book, I was beginning to feel that we humans had become so separate from the great animal kingdom that we were no longer much of a part of it.

Added to this is the modern tendency to isolate facts into small circles of specialization, and I can see why we often don't understand the environmental picture of life as a whole. Too much work regarding our mutual environment fails to consider the human animal as anything but a lesser part of our mutual world. This work does not.

In spite of the years of research that it took to produce this work, it's written especially for the average reader. Even though the subject is serious, I found it easy—very easy—to read. There is no lofty "I-know-something-you-don't-know" attitude anywhere in here.

Simply, this book is about the differences and (more importantly) the similarities between all members of the circle of life, very much including people. It shows clearly that all the members of all the different species have the same fundamental needs of air, water, food, and living space, a point I've repeatedly made in my own work. It shows where we *fit* or *don't fit* in the natural world, with an awesome sense of natural history about it.

Like my own films, this work does not necessarily divide entertainment and education; it's both at the same time. Passionately written, it becomes clear that we *are* a part of the circle of life, not nearly as separate as we believe. The author makes this point obvious by using facts we all know to arrive at a fresh perspective. The result is an exciting picture of the living world.

I envy you this fresh read. To my way of thinking, this is everything a book should be. This volume that you hold in your hands will encourage you to see all animals—even the human ones—in a different light for the rest of your life.

I'm sure you will enjoy this. You really will.

Marty Stouffer
Aspen, Colorado

Part I

INTRODUCTION
CONNECTIONS & DISCONNECTIONS

I now realize it had to be Bobo's thumbs that started it.

Among all the stirring sounds, astonishing smells, and exciting colors of my first trip to the zoo, it is Bobo's hands that I most vividly recall. They had a dramatic impact on my young mind.

Bobo was a massive chimpanzee, pacing back and forth in his cage, looking completely bored. As I watched him, dazzled by his enormity, I remembered the orange my mother had jammed into my pocket as my father and I were leaving home. It occurred to me that Bobo might like such a treat. I took the orange out of my pocket and, balancing it on the palm of my hand, my fingers held carefully flat as I had been taught to do when feeding an animal, I held it out toward Bobo.

Bobo's boredom vanished. His sudden intense focus on my extended hand was a little scary. The orange, bright in the sun, looked too far away, but I held my ground. I steadfastly watched my own hand as Bobo reached for the orange. His arm seemed unbelievably long. He plucked the orange from my hand with his *hand*— in precisely the way I would pick up an orange.

I hadn't expected that.

As I watched Bobo sit comfortably, chewing his/my orange, his hands held all my attention. Young as I was, I had already heard that the main difference between people and animals was *the opposable thumb*. Yet when I asked my father if Bobo was an animal or a person, he assured me that Bobo was definitely an animal. Puzzled, I watched Bobo's hands more closely; they were hairier than my hands, a little grubbier. His fingernails were chipped and irregular. But they were flat, just like mine. And Bobo used his thumb to balance the orange in his hand, just as I used my own thumb.

So what was the difference?

When asked, my father explained distractedly that scientists likely meant something more specific, something more precise, something more detailed than what I was seeing—a fine point about thumb differences that I would understand someday. It bothered me, but I kept my doubts to myself.

Returning to school in September, I asked Mrs. Seldon, the school librarian, about it. The book she gave me discussed how common the fully opposed thumb is in the animal world. Scientists knew perfectly well that the thumb was not a human distinction. In fact, one writer referred to the koala as the "two-thumbed teddy bear."

As life went on, the supposed distinction of the thumb as uniquely human died of natural causes, but it was quickly replaced with a slew of new absolutes: *Only humans are intelligent. Only humans have language. Only humans can see colors. Only humans keep animals. Only humans sleep on their backs. Only humans have a sense of time. Only humans are self-aware.* There are buckets of such absolute statements regarding what separates people from the rest of the animal kingdom.

In my younger days, I tried arguing with such comments. I quickly learned a painful lesson: I didn't know enough to argue intelligently. Early on, I began to perceive that people who made such pronouncements wanted their statements to be taken as absolute, unquestioned truth. The bottom line was always the same—there should be no question about human superiority, under any circumstance.

Yet if there were no genuine question about human superiority, why do such people always announce that *this* is the absolute difference? Eventually, I learned to listen in silence to such declarations. And I always believed I was ignoring their content. For too many years, I honestly believed I wasn't buying into the unquestioned superiority of humans.

—Norma A. Donner
August, 2002

CHAPTER I
THE ANIMAL KINGDOM

D ad was right. Someday I did understand the difference between human hands and Bobo's hands. And Dad was partly right about those differences, too—human hands are capable of more precise, more exacting, more delicate movement than were Bobo's hands. Bobo could never be a surgeon.

But who am I to sneer at Bobo? I can't be a surgeon either. First impressions count—Bobo's hands and mine are far more alike than different.

I had grown up understanding that scientists considered me to be a part of the animal kingdom. I had always felt comfortable as a part of the larger kingdom. Though I recognized that other people felt isolated by their separatist notions, I naively assumed I didn't have any of those.

Yet slowly and painfully, I came to discover that I did hold quite a few notions about my separateness. No matter what I had long believed, I did not miss out on the indoctrination of human superiority. My own convictions just weren't as obvious to me, at least not as obvious as others' convictions had been. Coming to understand where people do and don't fit into the animal kingdom has been, sadly, a series of disappointing discoveries of my own well-hidden convictions of human superiority. This realization didn't arrive as a great flash of understanding; instead, it arrived as a sense of constant leaking of those convictions. And let us not kid one another—not knowing they existed beforehand didn't make their loss any easier to accept.

I had written about animals for many years. I had even taught a college course on animals. I had spent my life living in close proximity with various other animals. I assumed I was at least beginning my trek from a better starting place than most people do.

I already knew that humanity belonged to only one of the three great divisions of nature—animal, vegetable, and mineral. While all animals, very much including the human ones, need elements of the other two kingdoms to survive, we people are classified into the animal division because we *belong* there. We are clearly neither vegetables or minerals.

Our inclination to deny or accept this classification has been on a roller coaster ride through time. The definition of *animal* has always reflected the median prejudices of the time it was written. As examples, here are some dictionary definitions:

> "Animal: any living being which has feeling and the power of voluntary motion as an insect, bird, reptile, or man or other mammal: distinguished from *plant*." [1926]

This definition sounds defensive. Why distinguish *man* in particular? Protozoa are at least as important as people and should certainly receive as distinct a mention.

Another example shows how these prejudices have altered over time.

> "Animal: a sentient living organism other than a plant." (Same dictionary: "Sentient: possessing powers of sense or sense perception.") [1956]

This definition seems evasive; it avoids the implied issue. The nineteen-fifties, when the definition was written, was memorable because it was a time of an overweening sense of scientific authority.

The definition of *animal* didn't change much until the early eighties.

"Animal: any living being capable of sensation and the power of voluntary motion; distinguished from *plants* by its inability to make its own food by photosynthesis." [1983]

This last definition strikes me as confrontational. We animals have slipped a bit by becoming distinguishable from plants through our *in*ability to produce our own food. While this definition is decently precise, it doesn't answer the question of *why* people are considered separate.

Beyond these primary definitions, all the dictionaries examined tended to separate people from the rest of the animal kingdom with a heavy sprinkling of words such as *higher*, *lower*, and *brute*.

Our individualistic, personal views on where people stand in relation to other sentient beings range between two extremes. One extreme claims to perceive no difference between other animals and human beings. Another extreme is that no comparison is possible between humans and any other beings.

Actually, these two positions have a history of their own. Looking back, comparing them with the rest of human history, no evidence exists that the earliest people believed themselves to be in the least separated from other animals. They knew perfectly well that they were just a part of the big picture of life. They could kill and eat, or be killed and eaten. With any luck, they could do the former and avoid the latter. It wasn't until people began to categorize other animals that the notion of separateness actually began to form.

Dr. Willy Ley, a naturalist, best known for his work on interplanetary travel, speculated that the categorizing of animals probably began long before written history. He suggested that primitive people had two simple classifications. They declared some animals *good*—that is, edible and not too dangerous to attack. They declared others *bad*—that is, dangerous to attack and not tasty. Dr. Ley extended this idea by suggesting that the earliest traveling people needed yet another category for animals—*new*. (This idea has the ring of common sense to it.)

The situation remained relatively static until Aristotle's time. It was this Greek philosopher who, in the fourth century B.C., said that science begins with marveling. His strong scientific leaning led him into investigating the three great kingdoms of animal, vegetable, and mineral. In fact, some historians credit Aristotle with setting up those classifications.

In modern times, it is difficult to understand why ancient Greeks didn't worry about being classified into the animal kingdom. The average citizen of that time was so disconnected from science that few were aware of the existence of the classifications themselves. From today's perspective, it's impossible to know how the people of more than two thousand years ago might have felt about this inclusion of people in the kingdom of animals. While it is possible that they might have been alarmed by it, it is equally possible that they might have thought it appropriate.

Aristotle was the first theorist about evolution. He speculated that life tends to advance from simple to more complex forms through time. And it was his sense of marveling that encouraged him to begin the arduous task of cataloguing known facts about animals. It was an enormous undertaking never before attempted.

It was a few hundred years later before anyone tried to add to this original work. The Roman naturalist Pliny the Elder, proceeded to increase the volume of known facts about animals by creating his own lists. By his time in the first century A.D., many more animals were known. Pliny's lists were heavily detailed, representing a huge body of work.

No one had yet attempted to classify animals in their relation to each other—a relationship that was heavily implied in Aristotle's earliest theory of evolution. Though other theories of evolution popped up occasionally, the relation of one species to another remained uninvestigated. Today, people perceive these two subjects—evolution and species' relationships—as integral parts, yet the two concepts remained separate for centuries.

It wasn't until Konrad von Gesner, a Swiss physician, published his first volume of *Historia Animalium* in 1551 that the cataloguing of animals proceeded. Dr. Gesner is considered the Father of Zoology. However, his work was little more than a continuation of earlier works. Dr. Gesner's lists were detailed, taking four volumes to complete. Nonetheless, they were just lists.

Meanwhile, the basic premise of human superiority was in full flower. How this idea developed is partly mysterious. Apparently, its justification seemed too obvious for anyone to record. And the conviction of its correctness continued unchallenged. It justified everything—cruelty for the sake of entertainment; the enslavement of other beings for work; the butchery of other animals for sport or food. Acting as self-appointed royalty, humans treated all other members of the animal kingdom as their subjects. The assumed inferiority of animals was justification for whatever humans felt like doing to them.

It was a long time before it occurred to anyone that the assumed inferiority of animals was a vital ingredient in our perceived superiority over them. Our recent ancestors even failed to notice this finer point. The assumption of superiority was by then so ingrained, the issue itself went unperceived.

What is too often forgotten is that scientific focus wasn't noticeably different from mainstream opinion. In spite of objectivity in the workplace, the people of science were, realistically, people first and scientists second. Armed with the very human sense of superiority, it was totally acceptable to subjugate any nonhuman to varying degrees of torture simply for the sake of scientific advancement. (The advancement was also assumed.) The distancing term *specimen* made it possible to make any dubious accomplishment in pursuit of knowledge. All experimentation with animals was automatically justified by that ever-present, subconscious conviction of innate superiority.

As we learned more and more about other species, the entire human focus remained adamantly anthropocentric. Animals were only analyzed and valued for the benefits they provided humans. At the time, it was considered proper and good to annihilate any species that did not serve a use to people. If any kind of being had the bad luck to be deemed harmful to people, or simply not directly needed, it often disappeared completely.

An unexpected and subtle turn in the prevailing attitude took place in the mid-1700s when Linnaeus, a Swedish naturalist, renewed the cataloguing of species. At this historical point, the study of animals was forever removed from amateur analysis to the domain of science.

Every naturalist before Linnaeus had tried to learn about animals through identification by their common names. Not only did these common names vary from nation to nation, sometimes they varied from one region to the next.

Linnaeus dispensed with this confusion by devising the double-name system, which is still in use today. His system delineated animals by genus, showing their relationship to one another, followed by species. As an example, the lion is classified as *Panthera leo*, and in this way is distinct from the tiger, *Panthera tigris*. This made it newly possible for a scientist from Germany (for example) to speak to a scientist from France in the new international language of science. For the first time each scientist could be certain that the other meant horse when speaking of *Equus caballus*.

In that one step, nonscientific people became cut off from the rest of the kingdom. To be sure, we were still absolutely certain of our inherent superiority. But all of a sudden, other animals belonged only to special people—scientists. And one result was the increase of scientific interest as well as an increase of evolutionary theories.

Linnaeus' naming system took animals away from the common people. Like throwing a stone into a clear pool of water, it continued to create ever-expanding ripples. This wasn't observed as very important to nonscientific people. It was just an elitist interest of scientific concern only.

While theories of evolution abounded, it was Lamarck's theory (Jean Lamarck, 1744–1829) that received the most public attention. In general, Lamarck felt that animals likely evolved to adapt themselves to their environment. And as long as theories of evolution—Lamarck's or anyone else's—were popularly perceived as of scientific concern only, no one was alarmed. Why should they be? It was absolutely understood that *animals* and *people* were two unassociated subjects.

Yet doubts and downright horror invaded the citizenry in the early eighteen hundreds when apes were first successfully shipped to the London Zoological Gardens. Comments about their form—"a parody of human shape"—created a good deal of disgust and disturbance among more sensitive men and women. Suddenly, all kinds of theories sprang up about a possible relation between people and animals. The very idea created shock and uproar—especially when mentioned in the same breath as those "disgusting apes," according to the writings of the time.

In 1931, an encyclopedia pointed out that science had already established the premise, that "only man and monkeys (including apes) have an instinctive fear of snakes." Unfortunately this absolute truth didn't hold up to scrutiny, even back then. Instead, it was discovered that young anthropoids (including human children) are taught to fear snakes; it isn't a natural fear at all.

That same encyclopedia reported that "the blood of man and that of any of the man-like apes are practically indistinguishable." This notion didn't hold up to scrutiny, either. (I feel obligated to mention the uproar about the baby who received a transplanted baboon heart.)

The more median view that developed between these two extremes was that, although man's body might actually be related to the other animals, his mind certainly was not! This remarkable conclusion seemed to eventually satisfy people of the time, and the controversy quieted.

But not, however, for long. The time for Charles Darwin had come. Born in 1809, the son of a prosperous doctor, he, too, studied medicine. Then he studied naturalism, but achieved no degree in either subject. He was eventually invited along on a trip made by the *Beagle*, a small sailing ship, as an unpaid, amateur naturalist. It was the mission of the *Beagle's* crew to survey the southern coast of Tierra del Fuego. Though the invitation was arranged through a variety of social connections, it was no small commitment for the young Mr. Darwin. The voyage lasted from 1831 to 1837. It wasn't until 1859, more than twenty years later, that Darwin published his thoughts on this trip in *The Origin of Species*. And then the uproar began anew.

One brief sentence in that work coalesced all the wrath that much of society felt toward science. It expressed itself against science in general, and Mr. Darwin in particular. All Darwin said was that "much light will be thrown on the origin of man and his history." That's it. He did not say, or even imply, that *man was descended from apes*, as he is often accused of having said. He didn't even think it.

What Darwin had implied (and believed) was that man and the other primates shared a common ancestor. Darwin believed in that mutual ancestor. And so all of the following fuss about the *missing link* was somewhat preposterous even then. No such being was ever found because it had never existed.

What Charles Darwin meant was that all primates had evolved away from that basic ancestor, just as much as humans had. All the primates were as different from that common ancestor as humans were. Because of the uproar that followed, this basic point is too often missed—evolution did not reach a pinnacle with humans and stop there. Evolution continued and still continues.

The entire fuss must be remembered in light of the times. That same 1931 encyclopedia laments about Darwin, "...a man whose daily life and character are a sufficient illustration of the impotence of 'Darwinism' to explain the extreme phenomena of Life and Mind...." In other words, the encyclopedia was saying that Darwin was considered a brilliant failure.

Meantime, as it became obvious to people that science disrespectfully classified humans as part of the animal kingdom, the social uproar escalated. From the scientific view, humans were unquestionably superior animals—but definitely animals—beyond a doubt. This view delighted some of our ancestors, the *no significant difference* group, but infuriated others, the *no comparison is possible* group.

In an effort toward fair play, the general trend on both sides was to demand that science focus on the issue of where humans were (and were not) separate from the rest of the kingdom—not arbitrarily, but scientifically.

This wasn't perceived by either side as a difficult goal. Some of our ancestors understood humans as abundantly different from animals. As a species, humans have accomplished more—far more—than any other kind of animal, and they wanted scientific acknowledgment. Meantime, the ancestors of others wanted science to prove that humans were the same, that there was no essential difference between us and all the other members of our kingdom.

The complexity of determining how we were different and how we were the same as other animate life turned out to be an overwhelming puzzle, unanticipated by the scientific community or anyone else. This issue arose a long time before scientists even recognized their own anthropocentrism—their focus on animals only as to how those animals might affect humans.

As a group, scientists never allowed the issue to be posed as a question. Not once. That the question itself existed, even within scientific circles, is implied by *answers*, never questions. It is only in examining how those answers change through time that I could see the question behind them. However, the answer of what separated people from the rest of the animal kingdom would not have varied through time, had there been no question. In the process of accumulating greater evidence that people were very much a part of the kingdom of animals, science disgusted the *no comparison possible* people, while delighting the *no substantial difference* group.

And each stated difference between people and other sentient creatures held an ethereal temporary position for awhile: a kind of, "Well, we thought humans were the only animals who used tools, but we've been discovering others...." Then, for awhile, "It seemed people were clearly the only smart animals, but recent discoveries in studies of whales demonstrate a need to update our understanding of intelligence...." Eventually, doubt of all the earlier simplistic answers came about. "We were pretty sure humans were the only life form that utilized language, but these dolphin studies raise questions about our current definition of language...."

As an issue, the similarities and differences between people and other perceptive beings is still very much alive. As yet, no one has actually determined if the position of *no substantial difference* or the position of *no possible comparison* is closer to the truth. Science gives distinctions on occasion, supporting one position or the other indirectly for a time. But because science is an ever-growing, changing body of knowledge, science cheerfully changes such distinctions once they have been shown to be false.

In recent years, the median position held that, while animals were nice, too much interest in animals indicated childlike inclinations. On the other side of the same coin, maturity was indicated by a mild interest in animals, with the understanding that a line could (and should) be drawn between people and animals. There was an implied understanding of where this line should be drawn.

It wasn't until well past the turn of the century that science began to recognize its own anthropocentricity. And with the beginnings of that enlightenment, it became clear that many former scientific views of animals were now questionable.

Through this closer scrutiny of earlier conclusions, science itself began to notice there might be a much bigger picture to consider, that the interaction of life in general might be incredibly important. One surprising result has been that science has been convincing people that we might actually need other animals—and that we certainly need them more than they need us.

Many people today grew up with this comfortable conviction firmly planted in their minds. As time passed, we have learned to question where we fit into the animal kingdom, but there is no longer any real question that humans belong to it.

Professionals in the varied fields involving animals had long been expressing concern about the issue of extinction before the general public began to pay attention. (Even the encyclopedia that pitied Charles Darwin's "failure," remarked that there was reason to be concerned that because apes were already so scarce it might be difficult to collect enough of them for human use.)

It wasn't until the first grave statistics about the whooping crane were issued that the median position regarding other animals began to shift dramatically. Never very numerous, for centuries the whooping crane was an irresistible target for marksmen. The bird stood more than five feet tall and flew with a wingspan of more than seven feet. Not only was it fired upon when in the air, the whooping crane also suffered continuous habitat disturbance on the ground.

Before I was born, there were only fifteen whoopers left in the main North American flock that normally migrated between Canada and Texas. (Another tiny flock, located in Louisiana, died out completely in 1948.)

Slowly at first, but continuously picking up speed, the median position began to shift to an awareness of the high number of extinctions in general. In addition to that appalling reality, the extinction *rate* was still accelerating. Entire species were disappearing.

In the last half century, environmental concern has taken on enormous importance. But the basic issue of how much like other animals we are, or how we are different, is still a foundation stone of this larger issue. Are we enough like other animals to need them for our own survival? Or are we different enough that we can survive without them? This is a vital point of the overall issue concerning our environment.

This relatively newfound environmental concern generally ignores traditional political, climatic, and geographic boundaries— as do the rest of the animals. It has quickly become a global concern, involving the global animal kingdom.

And this might be an exclusive awareness of humanity. This just might be a major difference between humans and other animals, referring exclusively to the awareness of the danger.

Without question, science was the instigation behind the concern with our global environment. Paradoxically, science had also been a key practitioner of the earlier anthropocentric view, which judged animals' significance only in relation to their usefulness to people. The latest emerging scientific view (not yet proven) is that life is a complex network unto itself, as represented by the biosphere. Grossly simplified, plants produce oxygen as a waste product; all animals consume this waste product and return carbon dioxide to the plants. The entire cycle repeats itself endlessly. Though this simplified version of reality has long been understood, it has only recently begun to dawn on people that, if we knock too many animals out of this cycle, plant life might also begin to die off for lack of carbon dioxide for photosynthesis.

Currently, while this is only a prospect, it is a real possibility. And the major result (from the human perspective) could be the end of life *as we know it*.

This prophecy of doom is but another anthropocentric view of the animal kingdom. (It simply questions how the extinction of a great number of other species will affect *people*.) However, it encourages a broader view of the importance of life as a whole. The saving grace of this approach is that it allows time for people to recognize that the similarities between all members of the animal kingdom are of infinitely greater consequence than any of the differences.

The shortsighted approach can be stated this way: "We have always freely destroyed other life forms; it follows that we have an established right to continue doing so. It will unquestionably be a shame, but not a big deal, if we use up all the animals. Their loss might result in some temporary confusion, but life will eventually continue as it always has."

Like it or not, this view superficially appears sound. People have gotten away with killing off entire species, often with science's blessing. And, as far as was understood only small consequences were perceived.

The less obvious effect was cumulative, perceivable over a long period of time. And that brings us to where we are now.

By this time, those people who have read or listened to an environmental approach to life at least suspect the assumptions of earlier times have been grossly inaccurate. If we do continue on, as we always have in the past, successfully exterminating many other species, intentionally or accidentally, we may wind up in far worse trouble than we can deal with. If we keep on "as we always have," life itself, as we currently know it, may come to a whimpering halt—for us all.

Being fundamentally logical, the majority of people want to know if such a worst-case scenario is likely before we exterminate too many other species. (How many is too many? We don't know.) And this is the most sobering issue involving people and other animals. Whether we like it or not, the fate of the rest of our kingdom could determine our own fate.

Even after centuries of earnest effort, we, as individuals, still tend to hold beliefs (as opposed to proofs) as to where we do and do not stand in the animate kingdom. The difficulty has been because, just as surely as we belong there, we also stand apart from the rest of its members.

We've set ourselves the difficult goal of carving an honest path through the varied beliefs of our kind, trying to honestly understand where we do and don't fit in with the rest of our kingdom's members. Were it not for our need for truth, we'd have long ago recorded in all the books, *people are the very best animals there are*, and we'd have left it at that.

After all, who is going to correct us? We write the books in our own languages; we read the books in those same languages; we distribute them only among our own kind.

But we haven't done that because we know perfectly well the question would remain. Today, many people have come to accept that we are just a part of the bigger picture of life, albeit an important part. And rather than the standard approach of *us* and *them* there's just *us* and that's fine. We're beginning to recognize that all our intellectual classifying and theorizing about our kingdom disinherited us from something spectacular—*and we want it back.*

It is only when we see ourselves as an integral part of this great kingdom that we can realize how much a part of our supposedly separate lives other animals really are. We are deeply involved with them and always have been. We rely on them throughout our daily lives whether we are aware of that reliance or not. Other animals have been an integral part of our history, even an integral part of our very civilization.

A good example would be the domestic cow. Even I was surprised at the incredible scope of her involvement with us. She has long been far more than milk in the refrigerator, meat on the table, or shoes for our feet. Around the world, her manure is a major source of fertilizer, but so is the rest of her body. And her fats and bones make many of the soaps with which we wash our faces and our clothes.

Historically, the cow was one of the main factors in the creation of urban centers. People could live in close quarters and still be fed from outside the urban centers by the cow. She is quite a remarkable being: she can consume incomplete proteins, in the form of grasses, and convert them to complete proteins, such as meat and milk.

The cow's involvement in medical history alone demonstrates well the subtle effects other animals have long had on our lives, both short and long term. The word *vaccine* is derived from the Latin *vacca* for *cow*. In the late seventeen hundreds, Dr. Edward Jenner noticed something peculiar about women who milked cows: those women seemed immune to the devastating outbreaks of smallpox. Eventually, he concluded that milkmaids were contracting a related disease, cowpox, from their charges. (Cowpox is a negligible disease in people—especially when compared to the deadly smallpox.) Dr. Jenner found that when small amounts of tissues taken from diseased cows were injected into healthy people, those people became immune to the lethal smallpox. And when Dr. Jenner published the result of his experiments in 1798, he literally altered the future of the entire human species.[1]

That discovery of long ago has developed an ironic twist in modern times.

Before specialization and growth exploded in agriculture, a farmer's cows might acquire a disease that could wipe out the entire herd. But the farmer still had other species of livestock to fall back on to keep the farm going. However, once monoculture livestock raising became the norm, this all changed. One of its side effects was concentration of disease. Maintaining only one kind of livestock could wipe out entire farm operations in short order.

While the specialization made it easier to attempt general disease control, it also tended to discourage the more personalized contact between the livestock manager and the animals being managed. (In simpler terms, the farmer no longer noticed if Bessie was off her feed.)

In an effort to keep disease outbreaks to a minimum, managers began feeding low levels of antibiotics to livestock. In the beginning, their goal was to halt outbreaks of disease. But it was discovered that most animals who were fed such mixtures had additional reactions: hogs gained weight more rapidly; cows produced more milk; and chickens laid more eggs. (Eventually, growth stimulants were also added to livestock feed in combination with the antibiotics.)

Recently, three forms of Salmonella bacteria common to cows became resistant to antibiotics. Predictably, these resistant bacteria can pass *from* the cows, *into* people via meat and milk. And because of their resistance to antibiotics, these three Salmonella bacteria (and possibly other kinds) have the potential to become a serious health threat to the human community.

Though the example used is specific, the potential problem exists with most kinds of bacteria. To the gratification of the *no substantial difference* group, many diseases are more generalized and may affect *mammals*. Both cows and people are mammals. So the potential for any disease to cross from cows (or hogs or sheep) into people is great.

Understanding all this, it's less surprising that Dr. Jenner's principle of vaccination seems to work in reverse as well. Bacteria may become immune to antibiotics by receiving small regular doses of similar antibiotics over a long period of time such as in cow feed.

I hedge here because none of this has been proven firmly either way. That surprised me. A few years back, the issue was reported at length in a variety of media. The best example was aired on the television news program 20/20:

> Dr. Scott Holmberg, Atlanta Centers for Disease Control: "As best we can determine, these drug-resistant strains seem to be causing more hospitalization of people who get affected by them, and more deaths from people who get infected by them."
>
> Fred Gutzman, American Cyanamid: "We believe that's absolutely false. We've seen no scientific evidence to indicate that there's any hazard to public health."

People may be growing taller because of growth stimulants fed to livestock animals. We may also be cultivating bacteria in our own bodies that are resistant to antibiotics. In relation to Dr. Jenner's earlier work, it gives pause for thought.

Becoming a vegetarian may help people avoid the impact of this group of facts about the cow, yet this animal is still practically inescapable in modern, civilized, everyday life. It is only one species and would probably pale to insignificance (in our daily lives) when directly compared to the laboratory rat.

I have yet to touch on the cow's overall effects on people. The numbers of cows are so high that they are considered major contributors to atmospheric levels of methane. The first awareness of this problem developed through a decade-long monitoring of atmospheric methane. At that time, it was learned that the increase in the cow population alone accounted for an 11 percent increase in atmospheric methane. Increasing numbers of cows on the South American continent is one of the main reasons for the rapid clearing of rainforests there. The constant grazing of cows on that continent keeps the land unforested. In the same vein, increasing numbers of cows on the African continent are thought to seriously contribute to the steady expansion of deserts, which in turn also affects global climate.

People are heavily involved with the rest of the animal kingdom. Human interdependence with other animals goes far and away past the casual in astonishing, complex ways. And yet, though we clearly interlock with the rest of the animals, we also really do stand apart from them.

Why?

During times of humanity's greatest steps forward, we developed the habit of patting ourselves on the back. Is there anyone left who doubts we no longer need to do that? Is there anyone left who feels that we still need to march along, proclaiming humanity as the best this and the finest that and always the most important whatever? I don't believe we're that insecure anymore.

I believe people are finally secure enough to honestly examine where we fit and don't fit into the most varied of the three great kingdoms. In my search for differences I found some very real ones, but always in unexpected places. And what I came to understand in the process of my search was an overview of the animal kingdom.

Within that overview was a surprising perception of the less tangible details of human relations with other animals. I can visit a zoo any time and find people held in deep fascination just watching other creatures go about their daily lives. Those people are watching for the simple pleasure of it. It doesn't seem to matter a whit to those enchanted observers if they know anything about whatever animal they are observing. Such people are enjoying a simple sense of awe and appreciation of the lusty richness of life. Their casual observations quicken their interest in everything alive.

Even the slightest passing thought about animals has a noticeable effect on the people involved. The good cheer of a person who has seen the first robin of spring radiates like a tangible vibration. This small incident can affect everyone that person meets during the day. In a subtle way, other animals tie our lives together from season to season, even from one generation to the next. Other life forms lend us a sense of the positive chronology of life in a way not equaled by clocks and calendars. On an atavistic level, I recognize spring and fall by the mass movements of migratory birds. I know it is summer when I see the scissor-tailed flycatcher visit my yard; I know it is winter when flocks of juncoes eat from my birdfeeder.

Some animals even tie the centuries together in some ethereal but deeply satisfying way. I only became aware of this a few years ago when I heard my first western meadowlark sing. Since the western and the eastern meadowlarks look so like each other, I was stunned to hear the thrilling unexpected melody of the western 'lark. When I looked him up in my bird identification book, I was drawn into a deep sense of affinity with John J. Audubon who had also been startled by the subtle difference between the two species. It was a marvelous feeling to experience a sense of kinship with another person who had lived and died long before I was even born.

The subtle effects other animals have on our lives are often felt most on the unconscious level. If we were unexpectedly faced with disposing of all our commercial amusements tomorrow and forced to spend our leisure time watching young animals simply learn the limitations and delights of their worlds, we would undoubtedly feel jarred at first by the adjustments necessary. But I'm sure in virtually no time at all our enthusiasm for life itself would be vastly elevated. No living being enjoys life more than young animals. Whether they are kittens or calves, goat kids or human kids, puppies or lambs, their glee and sense of awe is highly contagious. And all of us sober, serious adults are subject to infection by their sheer enthusiasm for life.

In nearly any circumstance where people come into contact with other animals—domestic or wild—it is always the people who get the most joy from the contact. Certainly, other animals often find such contact pleasurable, but the depth of feeling that people take away from such contacts is no less than magical.

Even those who never have much contact with other animals seem to feel it deeply important to know that they are there. I do not believe for one moment that any individual or group wants to lose all other animals, even if their personal involvement is only minimal or theoretical. I am rock certain of that.

The environmental movement of more recent times has done no more than strengthen the existing fundamental bond between us and other animals—that bond already existed in us all. Our concerns about our environment have strengthened this basic bond practically beyond belief and all around the world. There is no nation, rudimentary or advanced, that has yet failed to express an active concern about our global environment.

No matter which extremist view is involved, this bond exists for those people too. I've seen people who believe there is *no comparison possible* between themselves and other animals, swerve practically off the road to avoid hitting a cat. I've observed others, who believe there's *no substantial difference* among themselves and other animals, keep a dog safe on a leash during a walk. In both cases, interest in life is quickened in any relation with other beings.

Even those people who have no association with them declare readily enough that they like animals. A natural bond exists. As individuals, we may have no meaningful contact with them, yet still feel a need to know other animals are *there* even if we can't see or touch them. We may only occasionally pause in awe and watch the silent choreography of a flock of birds wheeling through the sky; we may never realize the sort of inter-communication that must be going on for them to fly in such perfect unison. It doesn't really matter; we enjoy a depth of satisfaction in the simple wonder of watching, a reward in itself.

Over time, I've seen a few people pause to listen to a singing bird then dismiss their own interest with a casual, "It's only a territorial declaration." Yet they still paused to listen, knowing very well the song was no less beautiful.

Even if you've never had much personal contact with other animals, you've likely wondered at times why we have slowly separated ourselves to an *us* and *them* status. Is such a separation is truly justified?

CHAPTER 2

CONCERNING INTELLIGENCE

*I*s intelligence unique to people? Until quite recent times, most folks believed that to be true. People were intelligent; animals were not. It was perceived as that simple. It was intelligence that made humans unique.

Intelligence should be easy to define. So what is it? A skill? Something that can be learned? Or is it simply a capacity?

Even dictionaries tend to define intelligence as an adjective. They usually describe it, more or less, as it is recognized, rather than state what it is. The implication is, *you'll know it when you see it*.

And that keeps the perception of intelligence a judgment call. When I meet someone, I can decide if that person is intelligent or not. The determination remains wholly at my discretion. Over the centuries, of course, there have been many sober attempts to get this determination out of the realm of personal judgment and more into the realm of objectivity. Science (for one) would vastly prefer intelligence to be something easily grasped and measured rather than something so elusive, so difficult to define, so evasive a concept.

It would seem that science itself, as a body of knowledge, failed to determine what intelligence was for a long time. Then it attempted to proceed on the pragmatic notion that it might work out the problem from a different perspective by trying to measure something it was unable to define. For awhile, this seemed like a pretty good idea.

Around 1800, the great French naturalist, Georges Cuvier, made the first official attempt to formally test for intelligence in other species. His test was based on ratios: the ratio of the brain's weight to body weight and the ratio of the brain's diameter in relation to the diameter of the spinal cord.

Cuvier's methodology, while creating arguments among his colleagues, satisfied most people for a while. There were scientists who liked it because it was a way of precisely measuring for the likely location of intelligence in other animals. Cuvier's idea also had the desirable quality of fitting in with the prejudices of the time. A few anatomists and physiologists eventually put Cuvier's theory to the test and it bit the dust: some individual dolphins and a few rodents (such as the rabbit) tested with Cuvier's method appeared to be as intelligent as people.

Of course, destroying Cuvier's theory meant it would be replaced. Other theories have come and gone: the complexity of the cortex (brain envelope), the size of cerebral hemispheres and fissuration of the cortex, density of neurons and cell differentiation, the ratio of body weight to brain weight—which seems barely different from Cuvier's original hypothesis—but this time, it eliminated measurements beyond the torso: species with tails no longer tested any better than tail-less species. (But I can see that the animal's shape demanded a judgment call by this method: at what point does a dolphin's tail begin?)

To list these attempts at how to possibly locate intelligence in other species without listing all the valid debates of the exceptions found to any particular rule, is clearly unfair to the scientists involved. In a cavalier way, it reduces vast amounts of work to a few paragraphs.

However, my goal here isn't to be fair (or unfair) to the people of science. My goal is to offer an insight into the pattern of the tests themselves: all of them use *Homo sapiens* for test design, no matter how tidily this fact may be buried under specifics. Not only are other species tested by human standards of physical criteria, they are also inevitably tested and compared to humans.

I believe this is fair. Every answer to every question must begin at a known point. Clearly, people are the one species we know to be intelligent and, logically, that is the right place to begin a search for intelligence. It is also obvious that any sort of intelligence that is unlike ours is not going to be found. (As example, it has been speculated that some insects have a sort of group intelligence and that would bear no relation to our sort of intelligence.)

Throughout history the search for a way to measure intelligence between one species and another has been under fire. Yet the tests for rating human intelligence, oddly, only came under fire in the 1960s. And it was the interpretation of test results that was criticized.

Sir Francis Galton, (incidentally a cousin of Charles Darwin) designed the first Intelligence Quotient Test, now more commonly called the I.Q. test, in the mid-1800s. His intent was to create a test that was not influenced by educational background.

Eventually, other tests were created to prove that people with better educations did indeed receive higher I.Q. scores. The Mensa Society, for example, is an organization whose members are in the top 3 percent on I.Q. tests. Yet when members of this group were given other sorts of tests, they were often found to score lower (sometimes much lower) than those people who did poorly on I.Q. tests.

Then there's the tacky little matter of gender. Throughout the rather long period when women were considered automatically less intelligent than their male counterparts, women kept testing about evenly with men. Many scientists of the sixties had the gall to openly question the quality of I.Q. tests on this point only.

The whole issue then took an unexpected turn, without any resolve of some of these earlier questions. Today, much focus is being placed on the difference between the brains of highly talented people and those of average intelligence. Current evidence suggests these differences are unexpectedly slight. Perhaps a thickened area over the auditory surface of the cortex (brain envelope) might have created a Beethoven or a Mozart. This is specifically suspected in Mozart's case, who became a virtuoso by age four. Current theory holds that such differences must be determined by nature (what we're born with) as opposed to nurture (how and in what surroundings we were raised). And yet, using Mozart as a good example of a possible weakness in this argument, it is relatively easy to conceive a situation that raises questions about this belief. Surely Mozart did not begin to play music before he heard it; clearly, he was first exposed to music—which would rest the case on nurture, rather than nature. And it might just be that unnoticed factors (such as the infant Mozart being deprived of tactile or visual stimuli) had a serious influence on his brain development.

Early into my research in this area, I could easily see that the whole issue of intelligence is muddied both by beliefs and differences in approach. And nowhere is the controversy more of a raging inferno than in the area of children. Stated succinctly, if humans are clearly superior, it logically follows that some sign of this imminent burst of intelligence will be discernable at an early age. I quickly learned that this isn't an area where it's wise to jump to conclusions.

Using most of the currently accepted measures of intelligence, chimpanzees don't score very well against humans. But they frequently surpass four-year-old human children. To exacerbate this point, chimpanzees also frequently surpass human adults on particular kinds of tests. Most commonly, these tests are based on ambiguity, for which more than one answer may be correct, and the animal—human or other—must determine the solution to the puzzle through logic.

The muddiness of this issue grows constantly deeper. When an American news series, "60 Minutes," checked into the backgrounds of Nobel Prize winners, it found that such people had very ordinary I.Q. test scores. And globally, such people are considered the very best of our species.

I've found it helpful to keep in mind that the issue of intelligence in other species has rarely been considered from a less than lordly level prior to the 1960s. It was only then that Dr. John Cunningham Lilly's work with bottle-nosed dolphins (*Tursiops truncatus*) became popularly known. And I was surprised to learn that Dr. Lilly chose to work with dolphins—not because there was existing evidence of their intelligence, but because of their size. Shallow-water dolphins were selected because they are not so large as to endanger the human testers and not so small (or rare) as to become endangered themselves.

It was one of the earliest and most disconcerting discoveries of Dr. Lilly's work that encouraged such a dramatic change in the median attitude toward other animals' intelligence. Unnervingly, Dr. Lilly and his colleagues were shocked to learn that the dolphins were also testing the human testers.

Within the group of dolphins being tested, it was noticed that the dolphins would emit sounds in a steadily rising pitch. The dolphins would stop raising the pitch when they were satisfied their human testers could no longer hear them. (This observation was verified through special, high frequency recording equipment.) Later, through the use of advanced recording equipment, the human testers learned they had also been tested (by the dolphins) to learn how low a sound humans could detect.

I can imagine how startling this discovery was to the human testers. Understandably, Dr. Lilly found it unsettling. And I've often wondered if the dolphins themselves were surprised to realize they were being tested for intelligence. (Why not? Everyone else involved was surprised.)

It was unquestionably Dr. Lilly's work that really flung open the gates of speculation on the innate intelligence of whales (*cetaceans*) in general. Prior to his work, few scientists had given whales any credit for intelligence. And it's still far from clear how intelligent any of the whales might be, though there's been a good deal of attention focused on them since Dr. Lilly's work. (It's not that anyone believed cetaceans were stupid or smart before. The entire concept of other intelligences was still fairly new.)

While there are always problems to overcome in testing any animal for innate intelligence, the problems are especially troublesome when it comes to the great whales. Their numbers have been so heavily decimated in recent centuries that extreme care must be exercised with the remaining animals. Oddly, there is far less danger for the (relatively) frail humans among these enormous animals; the whales are incredibly careful—in all circumstances—not to injure people. (It will be nice when the same is true in reverse; and it will come about eventually because of the enforcement of global laws designed to protect these behemoths.)

In testing for intelligence in whales, the problems of different atmospheres needs to be overcome: whales live in the sea; people live in the air. Not an insurmountable problem, surely. The tester can don diving gear. The reverse is not a possibility.

Having said that, I recognize it as overcoming only the simplest part of the problem. There are also vast differences in the two species' sensory equipment. Whales (generally) have poor eyesight; they depend far more on their ability to hear than we do. And it isn't known if deafness in a cetacean and blindness in a human are precisely equivalent. Certainly, this is one area of profound difference between the two species. How information is received by the brain may make all the difference in how it is interpreted by the receiver of either species. Our attempts to communicate with the larger whales may seem so alien to them that they fail to recognize what we are trying to do.

At present, it seems likely that the largest cetaceans may not even recognize *Homo sapiens* as any more than little animals that are sometimes amiable. Within the narrow areas that whales are exposed to us, our intelligence might seem questionable to them. Sure, we make the transition from the air (the atmosphere) to the water (the aquasphere) with mechanical aids. But whales see many other life forms make such transitions such as seals, turtles, and birds. There's no obvious reason for them to notice that we make such a transition only with mechanical aids. As far as we (so far) know, there is nothing in their existence that corresponds to our mechanically assisted breathing. In an effort to relate to what we actually do, it seems more likely they'd assume we're holding our breath, as they do, between surfacings.

Early on, I mentioned that there is also the problem of interpretation: at any point the interpretation between what was *transmitted* and what was *received* may become confused within the process of transmission/reception. Communications between people and other animals are in constant danger of either being misunderstood or simply ignored as the whales may be ignoring us. And the issue of trying to communicate with any other being is really only a side issue of the question of intelligence. (So far though, the issue of communication seems to be our only approach to the question of intelligence in cetaceans. This effort continues today as a direct result of Dr. Lilly's work.)

Still, there is at least one basic flaw built right into any system to determine if another species is intelligent or not. Normal scientific procedure insists that if a scientist in location *A* runs a set of intelligence tests on an animal, a scientist in location *B* or *C* should be able to exactly duplicate those tests and arrive at the same result. Superficially, this looks reasonable and fair.

But examining all the steps building to this surmise, there is an unjustified assumption that any individual of a given species is at precisely the same level of intelligence as any other individual of the same species. And yet our only certain example of an intelligent species—ourselves—can vary incredibly from one individual (a Nobel Prize winner) to another (a developmentally disabled person).

Understanding that this assumption is behind the standard scientific practice of exact duplication of tests with repeatable results, I can see that it ignores any possible variation in a species from one individual to the next. It also ignores the potential individuality in ability; it ignores possible past learning experience; perhaps most importantly, it ignores an individual's willingness to cooperate.

Even that apparently small issue of willingness to cooperate raises still another issue. It has long been suspected that science tends to grant higher intelligence ratings to those animals that will cooperate in tests. (Intelligence and willingness to cooperate do not necessarily go together—especially not when differing species are involved.) It has long been understood that even some human children score lower on intelligence tests simply because their attention wanders as they become bored. Anyone who has had even the slightest contact with another species knows that any animal may become bored with repetitive circumstances. Boredom is unlikely to be unique to humans—children or adults.

In this light, our old standard approach to other beings—the reward and punishment system—looks like it might be a lot less successful with many other animals. Some other beings may simply resent the system itself. And too often, unnoticed variables may alter test results. (Examples of such unnoticed variables might be hunger distracting the tested animal or sexual excitement stimulated by the smell of soap the tester had washed his face with that morning or some external distraction that the tested animal notices but which the tester is wholly unaware. These are simple possibilities that might have an undue influence on the test results.)

Generally speaking, there are all sorts of problems inherent in the area of testing for intelligence. While some of these problems are subtle, others are obvious. Many areas of doubt also exist about the interpretation of test results. Such problems may be no greater (or lesser) if we're only talking about the comparison of one person's intelligence to another's. But once the species barrier is crossed, any problem can become magnified enormously.

There is little doubt that the question of which other species are intelligent will remain a genuine puzzle for a long time. But while the above sorts of tests have been (and are being) tried, some entirely different approaches have thrown a different slant on the issue.

For quite awhile, the use of tools was believed to be uniquely human. Therefore, it was believed that tool-use might be considered a sign of intelligence. Scientifically defined, a tool is an article external to the body, used for a specific purpose and then discarded. In order for such a definition to be functional, it needs to be (and is) comparatively specific. An anteater that is using his tongue to retrieve ants from an anthill is not using a tool, as the tongue is a part of his body; neither is a bird lining a nest with feathers from her breast, since the feathers are a part of her body. Whether or not the twigs forming the nest technically fill the definition of tool-use is debatable.

There is, however, no debate about the chimpanzee that plucks and modifies a long stem of grass to poke into a termite hill as a means of extracting termites to eat: that grass is clearly used as a tool. And once such tool-use was openly accepted, other examples of tool-use in animal kingdom members seemed to suddenly crop up everywhere.[2]

There's been justifiable need for discussion on potential modification of the original definition of tool for a long time. For example, is it tool-use when a seagull carrying a shellfish drops it on a rocky surface to smash it for eating? Or are leaf-cutter ants using a form of tool when they chop leaves as fertilizer for the molds they grow as foods? It is partially because of these potential modifications of tool-use that it has occasionally been suggested that the word *tool* itself be redefined.

Our fluent use of language—in reading, writing, speaking, listening, and comprehending—has been considered a sign of our intelligence and a mark of intelligence in other species. But the whole issue of language and communication is so complex that it is handled separately in the following chapter.

There is, however, one point about language that needs mention here. The utilization of language is known to develop the brain of the language-user in a different (than any other) way. The left hemisphere of the brain of even a songbird is developed more fully than the left hemisphere of a non-song bird. Therefore, the species that utilize language have, in some real form, a different sort of brain entirely—even in the simplest structural ways. It may also be that the use of sound alters the brain, rather than the use of communication, as there is a noticeable difference in brain structure between those species who vocalize and those who do not. (If there is such a thing as telepathy, those beings who utilize it might show a different sort of enlargement or distortion in yet another part of the brain.)

Of course, there is also the pragmatic approach to the issue of intelligence. J. E. Lovelock, in *Gaia*, writes about the brains of whales: "...however it came about, the real point about the whale and the size of its brain is that large brains are almost certainly versatile."

No matter how much science would prefer it, intelligence is neither easily grasped nor easily measured. After extensive reading, I finally concluded that more than anything, intelligence seems to be a capacity. And the potential for that capacity is obviously not exclusively human.

In trying to examine intelligence as a subject unto itself, I came to realize that the apparent side issue of our concern for the well-being of our mutual environment shrieks of a formerly ignored symptom of intelligence: the capacity to be concerned for others, to extend concern beyond self. In itself, this may be a symptom of intelligence. In other words, concern beyond self may be an indicator of intelligence and intelligence may indicate concern beyond self.

In some ways, this line of thinking may represent a complete reversal of recent ideas. Just yesterday, the motto of civilized people everywhere was *Man Against Nature*. But only since we've begun searching for other intelligences have we been willing to state that nature itself is filled with wonder and delight. And slowly, we've begun to recognize that not only is man a part of nature, but the nature in man is also full of surprises.

Human development follows a standard progression, which everyone has observed repeatedly, no matter their circumstances. As an individual's intelligence develops, the individual's ability to be concerned develops on a parallel course throughout life.

When an infant is born, her personal comfort is the all-consuming desire; there is no reason for any other concern. If a baby is wet, hungry, or uncomfortable in any way, she will cry until the situation is changed. Being small and quite unable to do anything for herself, it's perfectly understandable that an infant demands a great deal of attention in her own self-interest. To call that interest *selfishness* would be accurate, but the word bears negative connotations in modern times so we avoid it.

Fairly early in life, the situation changes. Any infant begins to recognize other people as being directly important to him. Either parent, bending over the crib in a show of concern, will begin to gain the infant's attention. Once recognition of other people begins, the infant starts to be concerned about *who* is holding him.

Over time, an infant comes to recognize grandpa or grandma, brothers or sisters, friendly neighbors, more distant relations that show concern for her contentment. This represents the first development of any concern outside her own body. Though the number of people this child is concerned about will remain small for a time, it's still only the beginning of the natural twin developments of concern and intelligence.

As the child grows, he/she will suffer a sense of concern for anyone in this small group who is harmed in any way. They will defend these few people in whatever way presents itself. Sometimes, these defenses will be silly—"My Dad's bigger 'n your Dad"—but such defenses will always be in earnest.

As a child continues to move out into the world through play or school, she will be concerned about unrelated people: friends, schoolmates, and teachers. This represents the first development of concern about other people who may not be a part of her home life.

One of the most amazing things is possible—even likely—during this stage of development. In fact, it is so common, we hardly notice it. At any time during this stage of development, a child may begin to care about another kind of being entirely—a family pet—as a member of a select group worthy of her concern. (Sure, there are children who think of pets as objects and never any more than objects, but such children are a minority.) We tend not to notice this oddity of human behavior because it is so common, yet few other animals do anything even remotely like this at any time. (How many other species keep pets?)

As a child continues to grow, so his (or her) awareness continues to expand at a predictable rate, eventually incorporating his town, his state, and even his country. For a boy-child in particular, he may be expected to surrender his very life to prove his concern for his country, for his (presumed) dedication to a political ideology at an appallingly young age. Only recently has this natural expansion of concern for others received even a second glance by people. Typically, we didn't really note this feature of our own existence until we began to notice it in other species.

Nearly all animals react to signs of danger to one of their own kind. Even so, only some animals will actually offer genuine assistance to another (of their own kind) in a time of threat. And these animals tend to consistently fall into our consideration as the more intelligent species.

Swine will join forces to protect one of their number from danger. Swine have been known to successfully attack and drive away large predators who attempt to kill one of their group.

Wolves frequently risk their lives to save a member of their own pack.

Elephants are known to hold up an injured member of their group and assist the victim to safety. Elephants also act as midwives during birthing.

Dolphins, too, assist each other during birth. Dolphins are also noted for their practice of rushing to the aid of another dolphin to hold him up at the surface of the water so that he might not drown. Dolphins (as a group) will attack other animals that attack one of them.

Oddly, a dolphin that comes across a human floating in the water will push the human to shore. It is a clear effort on the part of the dolphin to do something beneficial for another kind of animal who may be in trouble. Upon hearing this, I heard someone remark that such a performance means little; a dolphin will readily do the same to a dead human floating in the water.

But so will we.

We know there is a possibility that a human floating in the water may still be alive, no matter if the person is conscious or not. We may not know how to go about resuscitating that human. Yet, if we possibly can, we will make every effort to get that human quickly to shore. We all recognize that there is little that can be done for a victim still in the water.

The desire to help another animal in trouble—especially one of our own kind, but often including other kinds—is a very strong, innate drive in human nature. And, while the desire is strongest when it is a matter concerning our own kind, we often rally (as individuals or groups) in going to the aid of other kinds of animals. At least hundreds of millions of dollars are spent around the globe each year to protect other animals. Only recently have people come to realize that other life-forms may be vital to human existence. People were concerned long before any of us realized the biological dangers of the destruction of other animate beings. For many people the bio-logical concern only reinforced their existing concern for the life-forms themselves.

Our tendency to be concerned about other life-forms is not unique to people, as the above-mentioned dolphins (who will push a floating human to shore) testify to their concern for another species. As mentioned earlier in this chapter, the care exercised by the larger whales to avoid injuring swimming humans is but one example.

Perhaps this has been a functional indicator of a way to locate intelligence in other beings right along. To a very real extent, the capacity to even be concerned about another form of life involves basic abilities that many animals simply don't have.

For example, an uncomplicated animal, such as a worm, may be quite incapable of the simple capacity to be concerned, even for its own safety. Crawling in the morning dew-covered grass, the worm may or may not be capable of even noticing a robin waiting to eat it. But the robin, in turn, is clearly capable of self-concern, for it will readily surrender the potential meal (of that worm) to fly away to protect itself from an approaching predator.

Taking a great leap in the scale of (apparent) intelligence: a wolf attacking a swine may get more than the meal he wanted when members of that swine's group hasten to defend the intended victim. It's also possible that the wolf will recognize this danger from its own experiences.

Philosopher Peter Singer has referred to this natural tendency as *the expanding circle of concern*. As I described earlier, a child quite natu-rally begins life as the center of his (or her) own circle of interest. The stages of growth as a person also include ever-widening circles representing first immediate family, then continuing to expand out-ward fairly quickly to even include the political division of nations.

Yet another use of the circle metaphor would be to use various animals to mark the ever-larger rings of concern, specifically as they relate to innate intelligence.

Using the worm as the center of the circle (with few or no personal concerns), as the concentric rings expand outward, we would next arrive at the robin (who is at least capable of self-preservation), and continue expanding concentrically outward to the wolf (who is capable of concern for other pack members). As the rings continue to expand outwardly, we would come to the dolphin (who is capable of concern for another kind of animal entirely). And at the outermost ring in this metaphor, we arrive at the apparent outermost circle, where there are many humans who are capable of concern for all other life-forms.

Perhaps this is only another version of the emotional human game of "Mirror, mirror on the wall, Who's the *smartest....*" One gloomy possible dismissal of this approach to the question of intelligence is that none of this can be measured in either millimeters or grams or even the ever-popular ratio.

In spite of our historical pride in the issue of intelligence, I believe many people (besides myself) are beginning to wonder why we should value any being solely on the basis of a single facet of his existence. Because of this, and because we now value other creatures for so many more reasons than ever before, the question of intelligence may eventually be dropped as meaningless.

As in every issue involving people and other animals, it will more likely continue to expand in unexpected directions. To date the only thing that is clear is that intelligence separates people from some of the other animals; people may be above-average in the animal kingdom in intelligence, but there's vast evidence that people aren't the only intelligent beings on this planet.

Human intelligence may be exceptional by degree in the larger animal kingdom, but not by kind. That much is clear. And, at this point, it's all that's clear.

CHAPTER 3

THE INFORMATION EXCHANGE

*I*t seems at first impossible to pry the subject of communication away from its complex attachments to the issue of intelligence. Yet once I succeeded in separating the two, I was rewarded by realizing this particular area of the larger issue (where people do and don't fit into the animal kingdom) held some unique information about people and other animals.

Looking into this, I began to subconsciously accumulate a list of questions I'd ask another species if I knew how:

- What do people look like from outside?
- Do you see our goals as worthy of our efforts, our sacrifices, and our dedication of our whole lives?
- Are we missing some major point that you perceive and we do not?
- Is there a bigger meaning to life that we've lost?
- Have people had genuine justification for the things we've done to you and your kind? To ourselves? To the biosphere?
- What are *your* goals? How do *you* see life? Is it all a mystery to you and your kind?

- Are we people on the right path? Do you and your kind follow another path?

As my list accumulated, I began to recognize my questions as originating from my most basic sense of humanity rather than from my political beliefs, my era of time, and my innate sense of human superiority. It also made it possible for me to perceive that the desire to communicate with other beings is the real reason behind the human search for other intelligent life, on or off Earth. People want to be able to communicate with another species; people want to be able to perceive humanity from an outside perspective; we people badly want to know if there is more to life in general.

The human approach to this goal hasn't always contained such poignant honesty. For (at least) centuries, interspecies communication was traditionally approached by looking askance at the rest of the animal kingdom—a kind of lord to slave approach: "If you can speak, speak up now or forever hold yourself quiet! *We* are the rulers of this kingdom—and don't you forget it!" Over time, the human position has softened somewhat to a more honest and polite approach. And one reason for that attitude change must be because the earlier approach was so resoundingly unsuccessful.

Yet even before people began to feel secure enough for politeness, there were a few who risked looking backward at how humanity got where it is in the area of communication. The study of languages is called philology, and the first philologists were brave (in their time) to examine our earliest developments in communication. In that time, the general conviction was that we people were born with all our wonderful systems of exchanging information just because we were so superior. But the pragmatic philologists noted that people taught children to speak so it all had to have started somewhere. And it certainly wasn't magic.

Try though they did, the early philologists could not prove the existence of writing farther back than about four thousand years. But they could prove that people had been talking to each other for about eighty thousand years. As one writer phrased it, ". . . the spoken word doesn't fossilize." Even so, other things do fossilize, such as human skulls.

Over time enough skulls were collected (from eighty thousand years ago), which showed alterations. These had developed directly from the human ability to speak: the jaw had become deeper, the tongue had become more refined, the larynx had begun to change—alterations that had improved *Neanderthal*'s ability to enunciate more clearly.

Humanity was chatting for a long time before it first occurred to anyone to try to communicate across any greater dimension than from one mouth to another's ear. Oddly enough, the first dimension that early people tried to cross was *time*. And it was only when I stepped back from this reality that I could see why this dimension (time) was tackled first. It probably began with relatively simple questions: "How can I tell my children (as yet unborn) how great and successful we all were working together on hunts? How can I tell my descendants, yet far away in time, that we lived this way? That we worshiped these gods? That this and that were the important features of our lives?"

Early people approached the problem in the simplest way possible. They used stories, often told in poetic form, passing from one generation to the next as oral history. And this method was remarkably successful for millennia. One of the oldest tales—*The Epic of Gilgamesh*—is estimated to originate from at least two thousand years before the birth of Christ. How this tale came to us from one generation to the next over thousands of years, is explained in one of many translations, here quoting N. K. Sandars:

> "What we have in both the Sumerian and Semitic versions {of the epic tale of Gilgamesh} is the word for word repetition of fairly long passages of narration or conversation, and of elaborate greeting formulae. These are familiar characteristics of oral poetry, tending to assist the task of the reciter, and also give satisfaction to the

audience. A demand for exact repetition of favorite and well-known passages is familiar to every nursery storyteller, along with the fierce disapproval of any deviation, however slight, from the words used when the story was told for the first time. Now, as then, an almost ritualistic exactitude is required of the reciter and storyteller."

Such legends have been successfully transferred through speech over impressive lengths of time. These old stories—eventually committed to clay tablets and from there to paper—succeeded in telling us much about how our ancestors lived before there was the written word. These ancient narratives have been transferred to us in modern times as epics, sagas, and myths.

Looking at this reality objectively, trying to shake my automatic notion that "we do such things better now," I was surprised to recognize that we still use oral history all the time in a variety of ways. Without oral history there would be no function to schools past the elementary level. Once children had learned to read, the more advanced grades would be superfluous and could easily be replaced by large book warehouses where children could simply pick up reading assignments. Unquestionably, the written word is a vital ingredient in our modern educational systems; but its use must be rounded through communication with another party—a teacher.

Oral history is today commonly used in subtle ways, but there are diverse forms currently in use. As example, I've heard several people comment that "parrots are only mimicking sounds when they talk and have no idea what they are saying." Yet, in essence, there is no essential difference between the parrot's talking and a person singing an advertising jingle. Or repeating a popular slogan. Or simply humming a song. Such things as platitudes, homilies, cliches, maxims, truisms, all serve to pass information (of a sort) from one generation to the next.

Sometimes slogans and apparently meaningless phrases are conveying subtle messages that aren't even necessarily apparent in what is being said. Stereotyping often passes from one generation to the next in this way, often without a thought, though some of the phrases used to accomplish this transfer of information sound as meaningless as the chatter of the parrot. One quick example would be the phrase, "they all look alike," referring to the Japanese people during World War II. What was said—"they all look alike"—vastly differed from the actual message—"they don't look like 'us.'" (It wasn't true, even then.)

Our basic animalness is clear in the way small differences in speech or pronunciation are sharply observed to alert individual listeners that the person speaking is an outsider. Differences in speech and pronunciation serve our strong sense of territoriality; it announces this or that class of education, this or that ethnic background, this or that income level. And this sort of subtle discrimination is most frequently used by observing the spoken word—our most ancient form of communication. It's nearly always observed most closely in a way that concentrates on differences, not similarities.

Looked at broadly this is by no means an exclusively human trait. All animals are constantly on alert for differences, however slight, in stranger animals. If bird *A* does not perform precisely according to bird *B*'s mating ritual, no breeding will take place. (That's why the Adelie, Emperor, and King penguins can all share the same rocky beach and not cross-breed.) In other species such subtle differences may rely more on other senses, such as sight and/or smell. The difference in humans is small and may not be exclusive to us: in spite of our varied ways of communicating, we tend to rely most heavily on our hereditary preference of sound.

One cannot examine only the finished product and understand the details of its development. This is as true of human communication skills as it is of any other product. And until we people recognize the developmental steps behind each of our own varied forms of communication, (how we developed them historically as well as how we develop their use individually) we will have no chance of recognizing any similar steps in other species. Without that awareness we are far less likely to learn to communicate across the boundary of speciation. We are also less likely to be able to observe the beginnings of communication in other animals or even to recognize the potentially diverse forms that may already exist.

Our ancestors developed forms of communication as the need for those forms developed. Once they had tackled the first dimension of concern—time—quite a lot of it passed before the next dimension—distance—was tried. And that dimension, too, was likely tackled for simple reasons: "How can I tell my neighbor that we will be hunting tomorrow? That we would like his help? How can I tell the people across the hill that we could use their assistance, too?" It was probably such simple concepts as these that started yet another sort of communication entirely.

These new forms of communication developed in varied ways, surprisingly dependent on location: the local landmass, of all things, was apparently a serious influence in the sort of system that developed. The Ashanti of Africa are accredited with the earliest known use of drums. For more than two thousand years they sent messages across dense jungle terrain using drums. Native Americans achieved something similar with their use of smoke signals across great open spans of ground. In roughly similar terrain, early Roman soldiers were known to send messages by flashing their shields in the sun.

For their smaller landmass, Canary Islanders developed a unique system of communicating over distance by whistling. The Incas of South America, faced with a vast and mountainous terrain, developed a system called the *quipus*. It involved tying a series of complex knots in the underfur of their pack animals (llamas) that could be read much like modern writing. Something similar was used in early Tibet to overcome their problems, though the animals used for the transfer of the messages were yaks.

It might have been as far back as pre-writing times that a peculiar distancing tendency developed between people and the rest of the animals. At the time, and even today, people relate to each other by gender delineation, often exclusively: "*She* did this" or "*He* did that." The distinction between one gender and the other is important to humanity, as it is everywhere else in the animal kingdom.

All animals seem to relate to the same or opposite gender with distinctive behavior. For example, the male European stickleback glows with color in the presence of another stickleback whether the other fish is male or female. But his behavior varies depending on the gender of the other stickleback: he will attack a male and court a female. The male wolf will not attack a female wolf, no matter the provocation. Cow elephants are intolerant of bull elephants at non-breeding times. And in relating from one gender to another, human behavior is most often as stereotypical as that of any other animal.

In language, however, people do distinguish themselves from the rest of their kingdom by a minor form of snobbery: people segregate other animals by denying them individuality even by gender. Most often, we refer to another kind of being as *it*, as if all non-human beings were machines or furniture—not animated life at all.

This common method of distancing people from other beings is notably like the oldest form of snobbery: the pronunciation game. *It* denotes an outsider by the simple expedient of denying personification by gender distinction. *It* also serves to deny any difference between one entity and another within a species: *it* can't be an Einstein or a developmentally disabled person; *it* is simply a dog.

Joseph Campbell was quoted on this subject in *The Power of Myth*:

> "You can address anything as a 'thou,' and if you do it, you can feel the change in your own psychology. The ego that sees a 'thou' is not the same ego that sees an 'it.'"

The average person may use *it* to identify a specific animal, but I suspect the most common reason for doing so is an unconscious way of hiding ignorance of the specific animal's gender. The distancing use of *it* is rarely used in nonscientific writing.

Personification is more important in most writing. But scientific writing is a whole different arena.

Clearly, I'm not referring here to *it* as an identification of an animal where the gender is not known; instead, I'm referring to scientific writing where the gender is an important factor in what is being said. I've been astonished to read such stuff often: "*It* gave birth..." or "*It* gathered its females together..." As a nonscientist this sort of vague language use strikes me as a sly way of retaining the old lord to slave approach to all other beings.

It also makes it clear that the written word can be (and is) used to carry various forms of bigotry, snobbery, and elitism from one generation to the next, often quite subtly by the simple use of word selection. Such attitudes were important to ancient humanity, so it really isn't too surprising that such messages were automatically carried on into the use of the written word.

When I look at the written word clinically, I can see that it succeeds in transcending the two major obstacles of other communication forms—time and distance—at once. Its creation unified humanity in a way that has yet to be surpassed by any other medium. As example, with either the medium of radio or of television I must absorb whatever information is offered at the rate it is given. With these two mediums, I cannot savor the words told me or weigh their impact when I receive them; I must accept information at the precise rate it's given to me or not accept it at all. But the written word gives me a choice in my absorption rate. (I can ponder or re-read what was printed.)

Years ago, the well-known astronomer Carl Sagan (1934–1996) refered to a library as a sort of communal memory. And it is in that sense that today, two thousand years after Plato's writing about his fears regarding the deforestation (and resulting erosion) of the Greek peninsula, I can read his work and share his fears about the same problems; the intervening millennia aren't a deterrent. Those thousands of intervening years are easily overcome by something as fragile, as delicate, and as insubstantial as a piece of paper.

In the same sense that a library serves as a sort of communal memory, it also serves as a time machine: the passionate words of Shakespeare, the intellectual arguments of Alexis de Tocqueville, the vicarious thrills of Mary Shelley's *Frankenstein*, and the morality plays of Aesop—all have the power to move me in one way or another, though centuries intervene between the writing and the reading.

From the apparently simple desire to communicate through time, then over distance, people have continued to improve communication systems. Many such systems developed through a lucky series of events, and surely, one logically led to the next. From each there developed what we now call spinoff ideas such as telegraphy, radio, eventually television, endless more. And each idea in communication began simply, demarcating a clear path to a more complex idea, which people automatically built upon.

Predictably, when it first occurred to people to try communicating with other animals, it seemed a simple goal. The oldest expression of that early idea is represented by the legend of King Solomon. At that early time, magic was a serious ruling force in the world, so that was how the good King Solomon solved the problem: he slipped a magic ring on his finger and then he could talk to all of the animals.

But that ancient and simple legend also demonstrates something subtler about the human community even that far back: that humanity already felt *cut off* from the rest of the animate kingdom. The legend shows that, there had developed a common belief that all of the animals spoke a common language that people had forgotten. It added to the human sense of separation. There's a special poignancy to that legend just because of that apparent side issue, a side issue that has reverberated down through human history, far past the time of the wise King Solomon. (It's also interesting to note that the king never passed along to the people what it was that the animals told him.)

This notion of communicating with animals via magic remained a set idea through much of human history. Nothing changed that approach until people hit the first gray area. Just a few centuries back, George Washington informed the Cherokee nation that its people would be used in an ongoing experiment to determine whether any Indian was educable or whether all the native people of the newly-formed United States would officially be treated as savages.

The fact that Indians could speak to each other impressed no one at the time. The Indians fell into that odd gray area: obviously not animals because they could speak; not necessarily people either because they couldn't read or write. (It wasn't considered relevant that most people in the world couldn't read or write—they weren't under Washington's jurisdiction.) Once again the issues of communication and intelligence crossed and tied themselves together. It was considered at the time that one proved the other—either way.

Some years into this experiment, a young half-breed named Sequo-yah (1766–1843) took it upon himself to try to create reading and writing for his people. I don't mean to imply that the young man was in the least concerned about Washington's edict. He didn't care a fig or a feather what any white man had to say about anything; he had grown up despising everything about whites. But he had noticed the white ability to communicate through the written word and that impressed him.

The young Sequo-yah had't learned to speak English. Then one day he came across a group of his people's children struggling to master a new language and a new skill—reading—at the same time. He was frustrated that whites should see his people as unintelligent just because they found it so difficult to learn two separate things at the same time. So he created a phonetic alphabet so that his people could learn to read and write in their own language—not anyone else's.

In every way Sequo-yah's achievement is one of the greatest in all of human history. He was the only individual to achieve the monumental task of creating an entire alphabet alone. He started and finished the monumental task on the simple knowledge that it had been done before. What he didn't know was that it had never been done by a single person before. So he did it. Alone.

I suppose I should eliminate Washington's pragmatic approach from the history of our desire to communicate with other animals. After all, it turned out that the Indians weren't simply other animals—they were people. And lots of people had said as much at the time. When it turned out to be true, some folks weren't all that surprised.

It was only a very short time ago that the magical connotations of interspecies communication shifted even a little. And that shift, though slight, was based on the pure and simple notion of human superiority. In 1920 Hugh Lofting, created the character of Dr. Doolittle who suggested that all people really had to do was want to communicate. His idea reflected the mood of the time—that people could do anything if they really wanted to do it.

Strangely, this tale demonstrates another major shift in our attitude toward other animals. Dr. Doolittle suggested that he could "talk to the animals" in their own languages: babble with baboons or chat with chickadees. This was a big change in our approach to the rest of our kingdom; quite simply, it demonstrates our awareness that there is probably no universal language between all the other animals—a solid idea from the past.

From this point, inevitably, some people did try to communicate across the speciation line. Most such efforts ended in failure and the few that were mildly successful received little publicity. (In the vein of Dr. Doolittle, maybe those early experimenters didn't believe hard enough.)

It wasn't until we began to investigate learning disabilities in our own kind that we took a second look at our own long-established communication methods. The four forms we most commonly use—reading and writing, speaking and listening—tie so neatly together that we often miss how distinctly different they are from each other.

Most of us take these variations in communicating skills as mundane, yet any *single* skill might be removed from the apparent whole and leave (at least the appearance of) the others intact. While I'm aware that one in five adults—forty million people—in the United States cannot read, I've also seen several such nonreading people copy a written message flawlessly.

Yet even the ability to see isn't a requirement of the ability to read. Many blind people read quite well but by a wholly different system than is used by the majority of readers around the world. And some blind people can even write quite well. Of course, those who are blind can speak and listen at least as well as any other person.

Some deaf people can lip-read so faultlessly that a casual observer may entirely miss noticing any difference. (I've had this happen to me, and I assure you, it's a profound shock to talk to someone for ten or fifteen minutes then discover accidentally that she is deaf.) Of course, deaf people who are in circumstances where they've been taught to read and write have no problems in that area or at least no more problems than anyone else.

It wasn't until I looked at how the majority of people learn to read and write that I began to perceive how mundane the process really is: we copy words. Copying written words or deciphering words that we read are essentially the same process. The alphabet itself becomes a Rosetta Stone to a new, infinitely larger world.

At some arbitrary point, different for each of us, seemingly of a sudden, *C-A-T* loses its individualistic meaning as a *C* and an *A* and a *T* and forms a conglomerate symbol for an animal I'm familiar with. It doesn't just come to stand for my personal cat, though she is in there too. Those individual squiggles (letters) come to stand as a single symbol for *all* cats. Suddenly, there's a larger meaning to the letters, a meaning that was hidden from me just a moment before, but will never be hidden from me again, and my world expands wildly through this relatively small experience.

Not all people learn such skills. The vast majority of humans around the world never learn to read or write. But that same majority learns to speak and listen. The abilities of reading and writing are therefore not necessary to learning to communicate. Conversely, since not every person can do all of these things, such things cannot be considered uniquely human. (If reading and writing and speaking and listening were uniquely human, all people could do all of them.)

Not much changed in the area of interspecies communication after Dr. Doolittle for quite a while. And then there was a sudden massive leap in the early 1960s when the work of Dr. John Cunningham Lilly became popularly known. It was his work that lit the fuse on all the issues that had grown up between people and other animals.

Before Dr. Lilly's work, the subject of interspecies communication had become isolated, reserved only for childlike delight through the work of Hugh Lofting. From Dr. Doolittle to Dr. Lilly was quite a jump. Suddenly a genuine scientific interest in the subject turned it into the heady topic of intellectual conversation. A well-established barrier crumbled without warning, unexpectedly opening over a vast chasm of new possibilities.

Dr. Lilly's goal was simple enough: to examine *Tursiops truncatus* (the Atlantic bottle-nosed dolphin) for language. Suddenly, humanity had a genuine scientist running a series of genuine experiments on something that had appeared magical just a short time before.

Everyone seemed eager to accept whatever might result from Dr. Lilly's studies. We would never have made that mental transition unless we'd felt we were ready. People seemed happy to welcome any new findings with open minds.

One of the milder, unexpected areas that opened up was a new field of entertainment: aquatic shows for public audiences. Cetaceans (whales) of all kinds were asked to cooperate with people: to jump through hoops on cue, to retrieve objects thrown into their pools, to leap into the air in response to a human signal. And to everyone's delight, this less scientific, more pragmatic approach to interspecies communication was, quite literally, a splashing success.

To be sure, much of the earliest contact between cetaceans and those people who hoped to train them to perform was conducted on a *me Tarzan, you Jane* level of communication. But those earliest trainers were very special people. They were confronted with the most basic questions:

- How does one go about requesting the cooperation of an alien being?
- How does one encourage such an animal to jump on cue? It is abundantly obvious that I cannot force a thirty-foot-long being [in some cases] to do much of anything, so 'request' is the right approach.

That such communication actually succeeded at this mundane level is clear at any show of this kind—and there are lots of them. Undoubtedly, how it began and succeeded so well largely falls under the heading of trade secrets. But even this seemingly mild area of interspecies communication had its hidden truths: trainers quickly reported that the old human training standard of *reward and punishment* was a complete failure. The sea mammals rapidly became bored and uncooperative with that method. But those early trainers quickly discovered that child psychology worked fine: reward good behavior and ignore bad behavior. (It generally works fine on people of all ages, too.)

Today, there is a well-developed respect apparent wherever people and cetaceans come into friendly contact. We people clearly admire these intelligent entities and are delighted that such feelings seem mutual.

Another subtle effect of these early successes: some areas where other animals have always communicated with one another stopped looking so alien to us. And that gave us a broader outlook when we reexamined the more mundane methods of communication apparent among other species.

We began to notice that there was no essential difference between the howl-along of the wolves and the sing-along of the people. The sense of group togetherness is the same no matter who is doing the howling.

Posture is another common method of communication in all species, though we didn't notice it much before. A submissive stance or an aggressive stance really aren't much different whether the stance involves two legs or four. And such mundane signs are usually easy to read from one species to the next. After all, wildebeest see lions all the time; they only begin to worry and mill about when the lion adopts an aggressive stance.

Animals that hunt in group formation (like the canines) must be communicating rapidly between each other via postural messages. Birds flying in large flocks tend to slow, bank and turn in formation, most frequently in perfect symmetry. We see other entities function in group situations in ways that indicate a certain amount of observance of postural communication, and we see examples of it all the time.

Large or small, groups of animals moving in a single direction attract others to do the same, perhaps simply by the powerful pull of their togetherness. And only some of this behavior can be explained away by genetically programmed behavioral tendencies. (There was no apparent genetic pull on the young people who flocked to Woodstock, New York, in the late 1960s; at least, most older people did not feel such a pull.) In migrating animals, there must be adjustments each year within the groups to accommodate changes as they crop up. And finally, it has only recently been acknowledged that many animals that migrate yearly do not take precisely the same routes from year to year during seasonal changes.

Mimetic (imitative) signs often cross the normal speciation boundaries. And people are as subject to these as any other animal. If I am standing in a group of people and one of them yawns, I usually find I've begun to feel tired. Some years ago, I discovered quite by accident that other animals suffer the same sort of contagious reaction to mimetic signs. Standing near a group of horses, I yawned several times, simply because I was tired. I was shaken into sudden alertness when I saw the first horse yawn, then another and another.

A sense of alarm readily passes from one animal to the next with no consideration of the speciation line. This is why many veterinarians move toward animals with such steady deliberation, only taking time for friendliness when they have completed the task needed.

Not all mimetic signs cross the boundaries between species, but they still pass readily from one individual to the next. Laughter creates the same sort of reaction in lots of people. Humor is pretty touchy when examined closely, but the desire to laugh must be an important part of what we laugh at. (If you aren't laughing, the subjective feeling that you are on the outside looking in is quite strong for many people.)

As always, it's children who shake the foundations of belief about the differences between people and the rest of the animal kingdom. Though an infant is not born with language, he (or she) is born with a natural tendency to develop the use of language eventually.

But for some months after birth, sounds themselves have no meaning to infants. It's still hotly debated as to what age an infant recognizes sound as possibly significant.

Yet it's long been established that it is necessary to speak to infants even before sound becomes significant to them; otherwise, the normal transition to their use of language might be delayed. It has been established that if an infant is not exposed to language through regular contact with people using it, the infant may never learn to use it for communication.

When a (human) parent is talking to a (human) infant, little of the one-way conversation actually involves communication except in one vital sense: concern is being transmitted. In a less tangible sense, familiarity with language is also being established. And eventually, familiarity with particular voices also becomes established.

Looking at this pragmatically, pretty much the same transaction is taking place between a pet owner and his (or her) charge. When I speak to my pet, I am transmitting concern, familiarity with the spoken language, and my particular voice. (These similarities are tacitly expressed by cartoonists who indicate the speech of a baby through the sketching of bubbles from a baby to the balloon of what is being said; cartoonists use the same system to illustrate what is being said by a pet.) Still another intangible result of speaking to an infant or a pet is that of encouraging personal bonding between the speaker and the listener whether that listener is an infant or a pet.

Of course, there is one important difference between speaking to an infant and speaking to a pet: eventually, the infant will actively utilize the language being spoken; the pet will only learn to utilize the language in a more passive, limited way—understanding that concern is being expressed, perhaps getting the gist of what is said. That last is no small accomplishment, but it's definitely limited.

Passively or otherwise, pets do learn to communicate adequately through using language, even within their physical limitations. If my dog becomes excited in response to the words *out* or *walk*, that suggests that she understands what those specific words mean. If your cat cries at the door to gain passage to the other side, he is demonstrating an understanding of *your* system of communication—and using it for his own purposes. There is no fundamental difference between such methods of communication and the more complex sort of communication between people except one: degree. People have larger vocabularies; we use communication to convey complex ideas to one another. But the essence of making another understand what is wanted is identical in all these cases.

There are instances where the essence of a communication can be quite different; nevertheless, a form of communication is taking place just the same. Done by humans, the *community announcement* is most often simple. Several easy examples that come to mind are a *No Trespassing* sign tacked up for public display; people wearing wedding rings transmit a similar general message. When a wolf marks his territorial boundaries with a sharply worded message in urine scent, he is also making a community announcement, leaving a message that only other wolves will heed. The wolf leaving the message doesn't care one way or the other if mice pay no attention; he doesn't mind if other mammals pay no attention; he won't even notice if birds pay no attention. His message is for wolves, and only for other wolves, very much the same as the *No Trespassing* sign is directed at people.

A bird singing his territorial song is making a slightly different sort of community announcement. Unlike the wolf, many birds don't even mind if other birds pay no attention to such announcements. The purpose of his message is a notification to an unknown bird of exactly his kind, and his kind only.

Some of our most basic uses of communication may differ immeasurably from all other animals in both essence and method. (Obviously not all, but definitely some.) And in that sense, many of our deepest convictions about communication may rest on subtle, exclusively human beliefs.

As example, take an elementary human subject: history. We cannot know if any other kind of animal has a concern about history, though it is very much an integral part of human life. Taking this issue on a theoretical basis, if there were such a thing as a whale historian (as just an example), what would he (or she) report to his kind? As far as we know, no other species practices war in our formidable style. Therefore, our theoretical whale historian would not likely have battle results to report, generals' names or war locations to memorize and tell his kind about. (And the loss of such material would thin our own history books dramatically.) The whole notion of history as a subject worthy of note may run deeper in human nature than we recognize.

Trying another example, as far as we've been able to determine so far, no other species dabbles in technology as we do constantly.

So if there is no war to report and no technological advances, why would there be any need for an historian?

Certainly, there is more to human history than just war and technological advances. But all of the basic concepts may simply be human in nature. Even the simple desire to be remembered, to stand out from the crowd, may be human in essence. As an historical subject, even art becomes eliminated since it automatically involves technology—which in turn may be a deeply human concept.

My own realization of this began (typically) from the opposite side, looking more closely at a kind of question I've heard all my life. As I was leaving a public meeting, an elderly lady was passing through the building's doors at the same time as I and I noticed her glance at the sky, then turn to her companion: "Do you know what star that is?" she asked. Her companion shook her head. The first woman looked up at the sky again and didn't seem able to spot the star she sought. "Oh well," she muttered, "I guess it was a plane. Funny, but I've wanted to be able to identify the stars all my life, and I've never gotten around to putting in the time to learn."

Driving home from the meeting, her comments kept running through my mind. After awhile, it occurred to me that I had heard the essence of her question repeatedly throughout my life: *Do you know what kind of animal that is? Do you know what that tree is? Do you know what kind of rock that is? Do you know what plant that is? Do you know what star that is?* Yet most of us can name all the larger wars, where they were fought, and which side won. I don't know what is more amazing here: how little we know about our everyday world, or how much we wonder about it without ever knowing. And, in turn, I find myself wondering if lions have names for the stars or if deer have names for the grasses, or if the simple act of naming things is, in itself, uniquely human. Sooner or later, humanity will be able to answer that question.

In recent times, we've done a great deal of research in interspecies communication. As a result, we may eventually know what other animals do and don't think about. Right now, we haven't even a decent guess. Yet since we've opened our minds to the possibility that other beings may be exchanging information with each other, we've begun noticing things that we just plain didn't notice in the recent past. While I doubt that anything we're noticing these days is actually new, the fact that we are noticing is what's new.

Evidence of existing communication between other animals has been accumulating thought-provoking evidence. As example, one widely reported incident that received a lot of attention a few years back involved killer whales (*Orcinus orca*).

In the Antarctic a band of several thousand killer whales began raiding the nets of a group of fishing boats. The captain radioed for assistance to a nearby whaling fleet, and the fleet sent three boats to help out.

It's important to understand that all of the fishing boats were converted World War II Corvette gunboats, identical to the three boats sent over (from the whaling fleet) to help the fishing fleet. Therefore, all the boats directly involved had the same hull design, the same engine sound, and even the same silhouette on the water's surface.

When the three boats (from the whaling fleet) arrived on the scene, one of them fired a harpoon gun and wounded or killed one whale. In less than thirty minutes, not one whale was near any of the three gunboats, though the entire group (of several thousand animals) continued to enthusiastically raid the fishnets around the other boats.

The only difference between the fishing boats and the gunboats was the small harpoon tripod on the gunboats' bows.

With several thousand whales involved, it is unlikely that they all saw the original shooting by just one boat. And none of the whales avoided only the boat that had fired the harpoon; they avoided all three boats with harpoon guns on their bows. And the whales didn't abandon the area; they simply avoided the potentially lethal boats.

This one incident raises complex and tricky questions: might the whale that had been injured noticed that the small tripod on the bow was directly responsible for his injury? And even if it was another whale that noticed it, what precise information was passed along to the rest of the group? And in what way was the information passed along?

This incident not only raises questions about cetacean communication, it also points out the deep changes in people's attitudes toward other animals. And those changes are dramatic. Not only was this complex incident noticed, it was reported again and again. Just a short time before, the same incident might have taken place and no one would have noticed. Just a few years earlier, the three boats from the whaling fleet might have arrived with harpoon blazing (or whatever harpoon guns do when rapidly fired) and whales would have been killed left and right without any thought involved by the people.

While this incident creates a sense of wonder about the possible complexities involved in cetacean communication, it also serves as an example of how people have changed their behavior toward other animals in a very short period of time. And this enormous change in human attitudes has continued onward.

In the 1960s R. Allen and Beatrice Gardner began to work on a different sort of interspecies communication project, involving an animal of an entirely different kind: a chimpanzee. And this chimp—Washoe—eventually became famous in her own right.

Both the Gardners are psychologists. Allen Gardner had already accomplished a great deal in experimental psychology with rats. Beatrice Gardner is also an ethologist—she studies animal behavior in natural situations as opposed to laboratory settings. This careful husband and wife team eliminated from consideration the idea of trying to teach Washoe their language; the chimpanzee just doesn't have the physical ability to enunciate human speech clearly. And since there is no known language among chimpanzees, the Gardners chose a median language, Ameslan (American Sign Language) which a chimp is able to speak as well as we are.

The choice of Ameslan had an inherent reward. Since no one involved in the experiment knew the language from early childhood, all the participants were, very roughly, on more equal footing.

It was typical of all such experiments, in that the animal involved had to be willing to learn what people wanted her to learn. Once again human standards, rules, and limits set the terms of inter-species communication. It is endlessly surprising how many other animals readily accept these terms and achieve fame with their success. In this sense, Washoe was just like all the others.

This well-known experiment created vast, raging arguments about phonemes, cheremes, exact word placement, and the possible significance of such word placement. It encouraged raging arguments on other issues as well, including what Washoe does and does not demonstrate, comprehend, and transmit. Undoubtedly, these arguments will continue, remaining highly technical in tone. But the fundamental concept—whether Washoe could or could not communicate—was proven beyond question through simplicity itself: when deaf people were shown films of Washoe speaking, they were astonished to see a chimpanzee communicating in their language.

The result of this one experiment also created, in many ways, a scientific revolution. And it continuously expanded its area of influence, far beyond its (seemingly small) area of scientific interest. Major new insights suddenly cropped up in theology and ethics, to name only two examples of seemingly disconnected subjects.

But the story of Washoe is hardly complete, even now. Today, Roger and Debbie Fouts are working at Central Washington University and are in charge of a young chimp named Loulis. Loulis demonstrated complete comprehension of more than fifty signs. That may not seem like much, since Washoe herself was able to demonstrate well over a hundred and fifty signs. The unique feature of Loulis' signing is that *a human has never taught him a sign.* What the Fouts did with Loulis was to turn him over to Washoe. And it was Washoe who taught Loulis everything he knows about signing.

The major achievements of such studies often appear small, but they reverberate continuously away from their origins, often seemingly far afield. Like everything else involving people and other animals, the communication issue continues to expand, often in unexpected directions.

Everything involving animals is now being observed a bit more closely with a fresh perspective. At times, putting old observations into new perspectives turns out to be rewarding in unexpected ways. As example, scientists had always noticed that elephants (both African and Asian) were able to maintain family cohesion, even when miles apart. But it wasn't until Katy Payne (who had spent much time studying whale communication) was invited to observe elephants that science learned how elephants maintain contact over distance. Dr. Lilly wasn't the only observer to be startled by different frequency communication between other animals. Dr. Payne observed the same sort of communication among elephants. Over long distances, they call to each other in extremely low frequencies, unheard by humans lacking special recording equipment. This discovery is recent, and it isn't yet known whether specific messages are being transmitted or whether this communication method is simply used for location.

What communication is (and is not) has far expanded beyond dictionary definitions. Such definitions reflect the old lord to slave approach between humanity and other perspective beings: "Communication: the act of imparting, conferring, or delivering from one to another; as the *communication* of knowledge, opinions, or facts."

A more satisfying definition of communication was suggested by Dr. Lilly in 1967, as "the exchange of information between two or more minds." By that definition, we can easily see that a good deal of communication takes place throughout our kingdom, from one singing bird to another, from the howling wolf to the deer, from me to my cat, and vice versa. The world we live in is abuzz with sound, and a good deal of communication is taking place through much of that sound.

I once listened to a physicist speak eloquently on our ability to wonder as uniquely human. His premise was that people are the only animals to wonder about things: what's on the stars, what's in the seas, how the universe works. I quite moved by his conceptualization, until I realized that people know so little about the other animals that such all-encompassing statements might represent our skillful use of language, largely to state unequivocally what we *don't* know. Eloquent or not (and he was), until we know a great deal more about what other beings do and do not think about, such comments are another way of patting ourselves on the back.

We already do that quite well. In recent times, on occasion, people have been surrendering their automatic sense of superiority, that sense of superiority that we've traditionally considered essential.

For many, though, that sense of superiority is being steadily replaced by a whole new set of inaccurate convictions. As example, it's now a commonly held belief that people have successfully crossed the line of speciation, that we can communicate with dolphins. Except in a limited sense, this isn't true. We still cannot answer those basic questions asked in the opening of this chapter: *What are your goals? Is life a mystery to you and your kind?* Unfortunately, this level of failing to communicate with other animals seems to be our failing—not theirs. We can't seem to ask questions in any way beyond the immediate, the mundane, and the practical. We haven't even been able to establish that dolphins actually have a language of their own.

On the flip side, however, it hasn't been proven that dolphins do not have a language of their own. Put simply, if we're so smart—and we clearly are—we should be able to resolve these questions. So far, we have not been able to do so.

There's work underway that's accumulating evidence that people aren't the only animals with the ability to "exchange information between two or more minds." The work involved is arduous, but progress is being made in colossal jumps just because of the new attitudes of people—that other beings just may be communicating.

In spite of all that, there still are some mightily unique features to communication between people. For one, people may be the only kind of animal that perceives all the rest in common, the only species that is concerned about possible communications with any or all of the others—not just the friendly ones or the antagonistic ones, not just the domestic and not just the wild, not just the dangerous or just the useful, but any and all of them.

A major ethical quagmire is developing where the study of communication between other animals is already underway. The issues involved are the same as those that crop up when thoroughly modern people locate a lost tribe of people who have had no contact with the world. How much should we interfere with their ongoing development?

Using chimpanzees as one example, quite a number of chimps readily communicate across the boundary of speciation now by using Ameslan. As in the case of Washoe's student, Loulis, chimps are clearly teaching each other how to communicate through a language that was given to them. We now know that they are intelligent beings, and we have given them a way to communicate with us.

On a profound level, why is it unethical to keep a human prisoner of war behind bars but ethically okay to keep an intelligent, communicating chimpanzee behind bars? And then, it follows that if it is agreed that our current treatment of chimpanzees in zoos or medical laboratories is unethical, how can we possibly continue? And how can we possibly draw the line at chimpanzees and not include the gorilla, which clearly has the same possibilities of communicating?

Then why not the dolphin? And why not the killer whale, who

clearly has the same capacities for communication?

The complications expand quickly. Why should we isolate selected species on the basis of only a single facet of their being—intelligence? Is that the only feature we should consider redeeming enough for which to offer protection? Should we wait until other beings develop a complex language of their own (presupposing they don't already have one) and then try to decode it? (That's something we've tried to do with cetaceans, and we've always failed.) Now we have the option of continuing along the lines of what we've been doing with chimpanzees—giving the gift of language to as many other animals as possible—as an easier solution.

If it is people who give the gift of language to other species, how different a reflection of humanity might we get from the gorilla, the whale, the dog, the lion, the anteater, or the zebra? It's very tempting.

What obligations does such a gift entail? Should rights be an automatic addition to the gift of language? (You can bet many people will fight that concept.) If people elect to grant simple, fundamental rights to another kind of being who is willing to cooperate in sharing information that we seek, what about those who aren't willing or cannot share information? And what about those who willingly tell us what we want to know: will it still be okay to slam their relatives into cages for whatever use we elect to put them? The ethical issues spread from this subject like ripples away from a stone thrown into a pool of water.

Aside from the ethical issues, there are apparently many uniquenesses to human communication. Some of them are subtle. I believe controversy is one of these subtle uniquenesses which is an advantage. Most people, I'm sure, would disagree with that statement instantaneously, but looking at it more closely, I think they would see that controversy is our best use of communication skills, and we use it all the time in that way. Without our tendency toward controversy, we people might never have succeeded as well as we have. If everyone always agreed with everyone else, human history would be one long record of placid and predictable events.

It isn't like that.

Controversy is an integral part of human history. Controversy has continually forced people to look things over in new ways from a fresh perspective, from a different view. Human communication skills make this advantage possible. People utilize impressive verbal and intellectual skills in controversial situations, more than at any other time. Watch and listen to any debate anywhere on any subject. It's easy to see that each side is forced to shore up points with new thoughts and original perspectives; stale concepts just don't work in controversial situations. The minds behind both sides of any controversy are working at maximum capacity.

I meet with a group of friends every few weeks. We all agree that debate is a mental stimulation missing in everyday life. We meet regularly, not to agree, but specifically to disagree in a tolerant, congenial, sympathetic atmosphere. (Any subject we all agree on gets dropped immediately as worthless.) Simply put, debate is enlivening; continuous agreement is dull. Debate makes for a roomier relationship, a more compelling friendship, uplifting conversation, a warm sense of fearlessness.

While it's rare for any of us to bring up specifically work-related topics at these meetings, I must confess that I brought up communication (in relation to this chapter) when I realized that one human uniqueness in the area of communication is war. Here we are, the species with truly magnificent communication skills, and we abandon them readily to toss bombs at each other.

Even those who deeply believe we aren't all that much different from other animals readily agreed that war is unique to humans. We sadly nodded our heads in unison about *man's inhumanity to man*, realizing there wasn't anything to discuss about it. Surely those who were convinced of human superiority would agree with this discouraging distinction.

One member of our group asked why territoriality was only horrid when people practiced it. "Why is it okay when other animals do it?" There it was. War in the way we practice it is just a difference in degree, not in kind, from the way other sentient life practices territoriality.

Controversy opened a new avenue of thinking. I found that once I focused on war in this way, lots of information supported that view. In 1693, William Penn "...pointed out that all wars arise from one of three causes: the desire to hold onto something that another wishes to take away; to get back something that has already been taken away; or to get something from another to which one has no right." One variation or another, the issue of war is inevitably territory, to control more of it than another individual or group—for whatever species is involved.

And since the winner of any such territorial dispute insists his (or her) version of the result of the war is the right one, the reason the war was fought in the first place will inevitably be buried under obscure justifications, piles of rationalizations, tons of defensive alibis. It then becomes impossible to know if life in general would have been better, happier, more stable had the other side won instead of the might makes right side—no different if this jackal or that jackal wins a dominance battle, no different if this marmot or that marmot wins a battle over territory, no different if this wombat or that wombat decides right through the use of might. There is no real difference here—not between people and other sentient life.

The only certainty in the area of communication is that people are *probably not* the only animals who exchange information between two or more minds. At the very most, the human difference in communication skills is in *degree*, not necessarily in *kind*. Whether those differences are as large as people believe them to be is still very much a question—like the questions this chapter opened with. Questions are what we have the most of, and there's an enormous number of people all around the world trying to achieve answers to those questions. It seems likely we'll get some of them answered, eventually. As long as a person is ignorant of the known facts involving the information exchange, it can be said that people are the only animals who truly communicate. But our incredibly skilled use of communication isn't the only area where we believe ourselves to be unique.

CHAPTER 4
OPTICAL ILLUSIONS

*M*ost of the research into animal vision is relatively new—and some of it is quite new. Vision has a distinctly startling relationship to a variety of unexpected subjects: the development of intelligence, language; the domestication process, and behavior. Oddly, the subject of vision doesn't always have much to do with eyes.

My first discovery was that animal vision was better understood on a pragmatic level in past centuries, far better than people understand it today. It was common knowledge among early falconers that on hunts birds could see farther and more clearly than people. Falconers often took along a shrike. The shrike would reliably clamor hysterically when he first spotted the returning falcon—long before the human hunters could see it. This little trick apparently made the waiting easier for the impatient hunters.

Falconers weren't the only group who understood that other animals' vision was different than human vision. Early whalers knew that most whales have a wide blind spot directly in front of them. Therefore, the wisest way to approach such behemoths was directly from the front. In that way, whalers could often get quite close to a whale before the whale even knew they were in the vicinity.

The average modern person understands less about an animal's visual ability than was understood in the past. For example, all my life I've heard that "only humans can see color." There is a riddle here that is unfathomable to me: why does the average person insist that other animals cannot perceive color? Especially when there is overwhelming evidence to the contrary?

The belief that "only humans can see color" is so widely spread, I assumed it came from scientific sources, but I could not nail it down. Still, it has become ingrained in human society and is accepted as an absolute truth. And with monotonous regularity, human decisions are based on this truth.

Some time ago, I read a newspaper article about a man charged with drunken driving in Louisville, Kentucky. The man claimed he couldn't be charged with drunken driving because, though he was definitely drunk at the time, he wasn't driving: his dog was.

I'd like to skip over the illogic of that to make another point. The man was charged and served time—not because it was an unlikely story—but because "...dogs are colorblind, and his dog would not have been able to distinguish between a red and green light."

Could it be true that "only humans can see color?" In pondering that, I wondered most how it is that other animals use color in so many pragmatic ways: *protective coloration*, or its variations, *camouflage coloring*, or *concealing coloration*. These things wouldn't work at all if other sentient beings couldn't detect any difference between one color and another. The lion (*Panthera leo*) in her tawny coloration blends perfectly with her native grasses; the Arctic fox (*Alopex lagopus*) would have no need to change color several times a year to blend better with the seasonal changes of his tundra background, if prey animals saw him only as gray. Coloration serves an everyday function for non-predacious animals as well: the bittern (*Botarius lentiginosus*) of North America freezes when approached, relying on her coloration to make her essentially invisible; she's even been seen to "...sway slowly from side to side, as if imitating waving reeds," normal background for a bittern. No matter what it's called, camouflage coloring works quite well throughout the animal kingdom from the grasshopper to the wolf.

Warning coloration works quite well, too. Bright and flashy colors are worn by a great number of animals to deter predators. The fire salamander of Europe (*Salamander maculosa*) goes through life in shiny black, mottled with bright yellow spots. The clashing stripes of many stinging insects work effectively to ward off predators (including people). Birds and monkeys have been noted to avoid the more brightly colored butterflies. And nearly every one (of any species) learns quickly that the more brightly colored snakes and frogs are best left completely alone.

Warning coloration is a good example of generalized communication across the boundary of speciation. Though every generation of animal must learn the hard way that bright coloration is a warning—*we do learn*. The warning of bright colors serves to announce possible bad taste, bad odor, or a hazard to your health. Fundamentally, warning coloration is quite successful.

Coloration has lots of positive aspects, too. A male cardinal (*Cardinalis cardinalis*) could simply be a dull gray if his mate couldn't see his flashy red. Fortunately, quite a few birds brighten gloomy winter weather, often with dramatically different coloration between genders. And their lustrous plumage makes the cold and gloomy landscape a brighter place. Logically, then, the world is a more colorful place just because coloration is so successful in so many living beings.

At this point in my research, I took my dog to her veterinarian and asked him how limited other animals' color vision might be. He stared at me, startled. Finally, he said, "I've no idea." He explained carefully that such knowledge had no practical application in field or office, so it's barely touched on in veterinary school. He also explained that when he had any difficult problem with animal vision, he contacted a specialist—a veterinary ophthalmologist. And he gave me the address of the one he used.

The first specialist I contacted recommended some books— which inevitably led to more books—which inevitably led to more veterinary ophthalmologists. I came to understand that vision in the animal kingdom is as varied as body structure. Contrary to popular opinion, there's no question among the specialists that animals perceive color.

And there was another realization attached to this that caught me by surprise: hidden in the belief that "only humans see color" is the subtle (inaccurate) assumption that humans perceive all colors perfectly. There isn't anyone who wouldn't quickly identify snowflakes as white—even though snowflakes are not white; they're clear. The same is true of the polar bear (*Ursinus maritimus*) who appears white to people, blending perfectly with his white snow background. What people are perceiving, when looking at either a polar bear or a snowscape is light deflection. And white isn't the only distortion of color that people see. The Indigo bunting (*Passerina cyanea*), appears living throughout the eastern North American continent, as a luminous, shiny, metallic blue. These cheery little bright-blue birds are actually black, only appearing blue to people because of the way sunlight breaks up when it strikes their feathers.

I became aware of this subtlety when I was casually watching an Indigo bunting sitting on a tree branch in the sun. When he flew out of the sun and directly into the shade, I was astonished: he literally vanished right in front of me. I was left watching a patch of shade with no fluttering wing movements. I was left watching nothing at all.

Like many of the other apes, people can distinguish most clearly between shades of green, which follows logically from our arboreal beginnings. But by no means does humanity hold any automatic superiority to all other animals in the area of vision and especially not color vision.

In their own unique (and glorious) way, flowers use bright colors much as birds do. The brighter their colors, the more insects they attract for the purpose of pollination. It's been suspected (and occasionally proven) that quite a number of insects that serve as flower pollinators can see a broader color spectrum than can people. And even the simplest life forms, such as animal protoplasm, can at least detect the presence or absence of light.

Color vision isn't even remarkable in underwater animals. Fish lures have always been made in a wide variety of the brightest colors because fish are attracted to such colorful objects. Recent research implies that such colors need to be very bright to overcome the dulling effects of deep water.

Broad groups of animals have unexpectedly exceptional vision. Amphibians must be able to see clearly in two very different atmospheres: in air, where light travels in straight lines; and in water, where light is polarized. In addition, a variety of fishing mammals such as seals, otters, and bears, must also have vision that functions in differing atmospheres.

The fishing birds, such as herons, have additional problems that they need to overcome. They must learn to judge the difference in visual information between two atmospheres. To avoid going hungry, fishing birds must learn that a fish appearing to be *here* is really *over there*.

Quite a variety of vision types aren't even vaguely analogous to human vision: with general vision in eyes roughly analogous to human eyes, the pit viper can also see into the infrared spectrum through pits alongside what I would call his nose. And this secondary vision is superb. Some of these animals can detect temperature changes about twice as fast as current instruments, even in complete darkness.

As scientists look further into other animals' vision, they find exceptional vision in unexpected places. In the dark depths of the oceans, there wasn't any expectation of finding good—much less exceptional—vision, but it's there. There isn't any reason (comprehensible to people) for squid to have exceptional vision, yet they do.

Long ago, I read a science fiction story that suggested a reason for the superb eyesight of the squid; that such visual acuity might be closely related to the squid's ability to change colors rapidly. The story suggested that a squid may communicate with another squid visually by rapid color changes in a sort of pictograph.

A more pragmatic explanation may be the squid's social system: squid swim in schools like flocks of birds. Their good vision may serve them simply by helping each animal to maintain his position and proper distance within the school. Yet squid vision is so well developed, so far ahead of apparent need, it leaves me wondering if the more complex idea isn't accurate—or at least a possibility.

Most of the vision elsewhere in the animal kingdom is more directly analogous to human vision, certainly more than it is to vision in insects, pit vipers, or squid. Most sorts of vision in the animal kingdom involve familiar technical features: involving things like cones, rods, photoreceptors, pupils, retinas, corneas, optic nerves, irises…just the same as ours does. But general animal vision does have some startling connections to odd subjects, like the development of intelligence and even the domestication process, as I mentioned earlier.

Some of the earliest research into animal vision started from the realization of how intertwined intelligence and vision seem to be. Examined closely, there's a direct correlation between the human sort of intelligence and the human sort of vision—which in turn relies on the use of hands (like human hands or Bobo's hands). Eventually, it was realized that three separate developments must occur *at the same time* in order for our human sort of intelligence to develop.

Using people as a familiar example, being able to simply walk without the need to use our hands freed those hands for other tasks—like eating. When the earliest prehumans could bring food up to their mouths with their hands, that left their eyes free to watch for danger while they were eating. Having both eyes on the same side of their heads gave them stereoscopic vision making it possible to judge distance, a feat which many other creatures cannot do. Probably sitting in a tree, this earliest prehuman ancestor could eat with his hands, leaving his eyes free to watch for any people-eaters in the area. In addition, his type of vision made it possible to determine that a predator was still far away so he might finish eating before he had to move along.

A good example of a dramatically different sort of vision would be any member of the rabbit family (*Leporidae*). Rabbit vision is both greater and lesser than human vision. It's greater in that a rabbit can see nearly all the way around himself, whereas people can only perceive (roughly) what is right in front of them. We people have fairly good peripheral vision, but not nearly as good as the peripheral vision of the rabbit.

Yet rabbit vision is lesser in a specific way: when a predator moves into a rabbit's vision, the rabbit cannot determine (visually) if the predator is very close or still far away. The human eye lens adjusts to accommodate a smaller focus. The rabbit's lens does not. He must sit there, quite still, unable to determine (again, visually) if the predator is right on top of him or still far away. The rabbit sits and observes the surrounding scenery and the predator, as if all he could see was of equal importance; only experience and his ability to detect scent give him any clue about the best time to flee. While rabbit eyes look very much like human eyes, they function quite differently. Because the rabbit must put his head down to the ground to eat, his nearly 360-degree vision becomes a vital ingredient in his day-to-day survival. But it's learning from experience, more than his eyesight, that makes his day-to-day survival more likely.

Though rabbit eyes and human eyes look quite alike, the similarity of the appearance of any animal's eye to another animal's eye

may not be meaningful. On the other side of the same coin, however, differences in appearance between one kind of eye and another
may not be relevant either.

One glaring difference between animal eyes and human eyes shows
up in car headlights. While driving my car at night, I certainly
noticed that other eyes shone back at me in my headlights. And only
one animal I spotted didn't have brightly glowing eyes: he was a
young man walking along the shoulder of a rural road.

The whole thing puzzled me so I dropped into an optometrist's
office the next afternoon. He told me cheerfully that "all animals
have 'eyeshine.'" Not wishing to start an argument, I didn't mention
the obvious exception: the young man's eyes that I had seen. By this
time, I'd established that the absolute, "only humans can see color,"
should be regarded as a clear case of *Human Superiority Complex*. And
I'd already begun to suspect this optometrist's information was yet
another layer on the onion of human beliefs about other animals'
vision.

That night I drove around some more. I passed a domestic cat
walking along the edge of the road; his eyes shone back so brightly,
they seemed to be lit from within. As I drove around a bend, a man,
surprised by my headlights, glanced hastily away. But for an instant
there, his eyes had flashed a little—just a little. Over a week of
cruising around in the dark, I saw cats, dogs, raccoons, opossums,
and deer, and each had eyes that shone back at me. I remembered the
man's eyes had only flashed briefly, and the flash had been dull in
comparison to all the others.

I returned to a veterinary ophthalmologist to get straight on
this. The optometrist I'd spoken to earlier had referred to this glowing light as eyeshine. The ophthalmologist explained that eyeshine is
caused by a metal protein layer of reflective cells located behind the
retina in some animals' eyes. It's called a tapetum (pronounced ta-
PEET-um). Nocturnal animals (out and about after dark) and many

crepuscular animals (those who are normally out at dawn and dusk) have a tapetum in their eyes. That's the tapetum's function: to magnify small amounts of light, making it possible for nocturnal animals and crepuscular animals to go about their business in less light than day (diurnal) animals need.

Diurnal animals, including people, don't generally have a tapetum. My local optometrist believes all animals have a tapetum because those are the animals he's seen after dark. He has never seen a hog at night; he has never seen a songbird in the dark; he has never noticed a gorilla's eyes on a late evening—lucky man. And the ophthalmologist explained that the young man walking on the side of the road with the dully flashing eyes had only been caught by surprise with his pupils wide open. The mild reflection from his eyes had been from his plain old diurnal retina.

For a time, I continued to drive around at night because there was something else that had piqued my curiosity. No matter how much I read on the subject, I could find little scientific explanation for why an alarming number of other animals seemed to make an effort to jump right in front of my car. I was distressed at how often I had to slam on my brakes. (After all, I was driving to observe— not kill.) It puzzled me why so many animals seemed to wait until I was nearly parallel to them before they jumped out of cover, hitting their best speed as soon as they were directly in front of my car.

It seemed an insoluble enigma until I tried to see what was happening from a pragmatic perspective. As I drive near them, what is it those beings perceive as happening? If it were me standing by the side of the road, trying to work up my courage to cross that empty expanse with no cover, what would make me run out the instant I saw...what?

It struck me then that maybe they were perceiving a predatory beast, large and weighty, with huge glowing eyes, headed straight for them. Such a beast coming straight at me (for all I could tell) would probably frighten me into top speed instantly. I would have no way to know this huge predator couldn't see me, hiding there on the edge of the clearing. As well as I might be hidden by brush or scrub, I would be able to see its glowing eyes bearing down on me. There would be nothing in my inherited instincts to help me understand the beast was not directly pursuing me; there would be no reason for me to understand it wouldn't swerve and pursue me into the woods. Pragmatic as this notion is, it strikes me as a reasonable explanation for this odd behavioral tendency.

Birds show something similar to this behavior when they are suddenly presented with the eyespots on the wings of butterflies or moths. When a bird locates a resting butterfly and approaches with the idea of eating the insect, the butterfly flicks his wings open to reveal the eyespots. The bird, suddenly confronted with a large pair of predatory eyes looking straight at him, hastily flies away.

Generalizing greatly for simplification, eyes that have good nocturnal vision have a higher number of rods; eyes that have good diurnal vision have a greater number of cones.[3] Still generalizing, those animals with a tapetum in their eyes have a greater number of rods and a lesser number of cones. That should mean that those animals with tapetums do not perceive color well.

However, those animals with tapetums often do not fit neatly into a category; there are many, many exceptions to the above generality. This is the main point where commonly held optical illusions cross: many other animals, besides people, lack eyeshine. But there are also many other species that perceive color quite well—even with eyeshine.

One of the better-known exceptions would be the domestic cat. Again and again, the domestic cat is an exception to the general rule regarding eyesight. One of the veterinary ophthalmologists I spoke to told me quite calmly that his own research had convinced him that domestic cats can discern about a third more colors than people can detect. Other veterinary ophthalmologists disagreed with him, but readily agreed that cats do have good color vision.

Yet cats have a tapetum and they can also see quite well in much lower light levels than people can. One possible explanation for this is mentioned in Gordon Walls' *The Vertebrate Eye and Its Adaptive Radiation*. Dr. Walls explains that, normally, the difference between rods and cones within an eye are quite distinct, but such differences may be muddied by recent evolutionary changes. It seems likely that this is the case with the domestic cat.

Oddly enough, a similar sort of change takes place during the process of domestication. By far, most domesticated animals are nearsighted; that is, they can see things most clearly when those things are near them.

This alteration of vision through the process of domestication is most clearly shown by those species where only some of the animals in a given species have been domesticated. As an example, domestic sheep tend to be nearsighted, while their wild relatives retain normal—or even farsighted—vision.

By the time I'd gotten this far into the subject, I wasn't too surprised to learn the domestic cat was once again an exception. Long before, I had noticed that domestic cats seem to eat everything in a grasshopperlike squatting position. Even eating from a plate, my cat tucks her front feet under the rim of the plate. To her delight, I spent a few days studying her eating style by sitting on the floor and watching her while she ate. I noticed that she seemed to locate dropped morsels by scent or touch. And she never seemed to be looking at what she ate; her eyes held a dreamy, unfocused expression until she'd completed her meal.

According to Sandra Sinclair in *How Animals See*, cats can see best at distances of six to eighteen feet. Apparently, cats cannot see clearly, if at all, when objects are closer to them—unlike most domesticated animals. Watching my cat hunt, I can observe that six to eighteen feet is generally the distance involved in the average tackling of a mouse. Once her prey is closer than that, she simply hangs on, keeping in contact through her other senses: touch, sound, and scent.

Yet another layer of belief we people hold is the subject of weeping. Everybody seems to have feelings on this subject. Even many scientists state hotly that "only humans weep tears." Yet some people are convinced that animals weep tears, much as humans do.

I found it difficult to sort out the strong speeches of either side on this issue until I recognized that it was just a technical version of the *no comparison is possible* school and the *no substantial difference* group. The key ingredient in both arguments in this version is emotionalism—not just by the people doing the arguing, but by the issue itself. The actual contested point isn't related to tears at all, but to the question of whether or not animals can suffer emotionally. As long as you leave the actual production of saline tears out of the arguing, these two apparently differing groups could probably agree with one another.

Many animals shed tears for eye socket lubrication. Few actually shed tears for the reasons people do. I hasten to add that some animals may die from emotional distress—nearly everyone has heard at least one story about a dog who refused to eat after his master died until the dog proved beyond question that his emotional suffering was genuine by dying. But that dog did not have the physical equipment to lessen his distress through the release of weeping. Instead, he died, proving beyond doubt that his distress was both terrific and pure.

That dog shed no tears—not the way people do. The key word here is saline—not tears. Weeping saline tears (as people do) is a trait found only among those animals in long and close association with the sea. While the arrangement of tear-producing glands vary from species to species, only those other animals in close association with the sea—currently or in the recent past, actually weep, as people do. Otters, sea gulls, surprisingly elephants, and many more animals weep saline tears in response to emotional distress.

According to Elaine Morgan in *The Descent of Woman*, weeping saline tears in reaction to emotional distress is the human way of shedding excess salt from the body through the lachrymal glands. We people do this through our eyes, as do many other mammals. Some birds do this through their beaks.

The equipment for weeping varies from one animal to another. The ability to weep is not unique to people but, over the entire animal kingdom, it is uncommon. Many animals simply don't have the necessary arrangement of glands to produce tears for anything more than simple eye socket lubrication. Clearly, this has little to do with emotional distress.

What it has a good deal more to do with is Sir Alister Hardy's theory of human origins. In 1960, Professor Hardy suggested that many of our odder features—the sensitivity of our fingertips, our ability to weep, our arrangement of body fats—can be explained in terms of an aquatic life being a substantial part of our recent evolution. A better-known example of his theory is the recently much-touted *dive reflex.*[4]

Finally, I had reached the center of the onion of optical illusions: *the power of the human stare.* I was able to trace this idea to *The Jungle Books* by Rudyard Kipling. I had a difficult time with this because I had accepted it without question, as a child. I found it difficult to look at it again objectively with an adult perspective.

If I remember right, Mowgli discovered the power of the human stare when staring at wolves around the Council Rock. Mowgli frequently enjoyed making the other animals at those meetings turn away from his powerful stare. Rudyard Kipling also noted that the other animals never quite trusted Mowgli, just because he stared at them all the time.

My own powerful stare works well on my dog. But there's a possible reversal in there as well. If I drive through a safari-type zoo and stop to look at a lion—if that lion suddenly looks up at me, focuses on me, I will instinctively break the eye contact to look over all possible openings in the car. And if the lion continues to stare at me, I'm certain I'll suddenly remember someplace else I should go.

If the identical situation occurs when my dog is in the car, she will meet the lion's stare and return it, boldly. I know her well. She has complete confidence that my car will protect her. (I've already admitted that I don't really share her confidence in the car's protection.)

Is this an acknowledgment on my part that the lion is a more powerful being than I? On the opposite side, does my dog really believe she is a more powerful being than that lion? (Or is it simply that she has greater faith in the car?) Intellectually, I know perfectly well that my dog is right; it is highly unlikely that a lion can break into my car. But I'm still unsettled by the lion's stare.

My dog's reason for turning away from my stare is explained by Dr. Konrad Lorenz in *Man Meets Dog*. The eye structure of primates is generally different from the eye structure of most mammals. Many mammals have a wider range of vision than primates; they can see with their peripheral (side) vision, more clearly than people can. Because of this, there is usually little reason for most mammals to look directly at people; there is no need. People, on the other hand, must focus directly on objects and animals in order to see them clearly.

Consequently, most other beings find the human stare totally unsettling. There isn't any circumstance in their daily world where a stare bodes anything less than trouble.

Ultimately, staring is an aggressive act. On an instinctual level, other animals equate a stare with being the focus of an aggressor— most often, by a predator. While my dog is a predator, so am I. Yet we are both unsettled by a stare: she by *my* stare and I by the *lion's* stare. And while any person staring at her, no matter who they might be, unsettles my dog, she is most unsettled by my stare. Conversely, I don't feel particularly confident if she stares back at me.

But then I don't like it if anyone stares at me. And there's an opposite possible scenario in there, too. If I'm standing in the (inevitable) line at my local post office, and I casually let my eyes wander around the room, deep in thought, all is well. But if some thought distracts me while my vision is roaming, my eyes will fall at random at some point away from me. As my eyes gradually refocus, if there is another person standing within my range of vision, I will inevitably notice a frown, perhaps even a glare, in the other person's eyes. Realizing that I've committed a social no-no, I will look quickly away. We all do this sort of thing ten times a day and barely think about it.

Yet there's a good deal going on in these common exchanges, it seems to me. What makes the stranger (whom I have accidentally focused upon) look around in the first place? What primitive protection device is triggered? I've tried to figure this out from a more passive position (of allowing myself to be stared at) from time to time. All I feel is a sudden awareness that I'm being stared at, a rush of discomfort, and a strong desire to stare back. On the occasions when I deliberately gave in and stared back, most strangers looked away instantly.

Physical nearness seems to be an essential ingredient in the success of this interchange. If I stare at someone far away, they rarely notice. But if I stare at someone close to me, the reaction is practically instantaneous.

Simply put, *staring is an act of aggression*, fundamentally and finally. It is an act of aggression when I stare at my dog. It is aggression that inspires the lion to stare at me. It is aggressive behavior when I stare at a stranger in the post office—and it is a war of aggression if that person returns my stare. In every case, it is aggressive intent that's being communicated from one individual to another. And even across the boundary of speciation, the message is loud and clear.

Is the power of the human stare effective? Like Mowgli's wolves, my dog is a social animal. She recognizes my stare as an act of aggression. However, she knows me well enough to give me the benefit of the doubt; she will simply look away for a time when I'm staring at her. If my stare continues beyond any obvious reason, she will find someplace else to go.

Like her, I avoid people who make me feel uncomfortable. Also like her, I'm a social animal who feels the aggression of the stare. So apparently the power of the human stare works on me, too. In the United States Armed Forces basic training, men are taught to avoid making eye contact with an enemy; my spouse tells me he was instructed to avoid eye contact at all cost, to look at the interrogator's forehead, rather than his eyes, where such an evasion (of eye contact) can't be detected. In itself, this instruction is recognition of the power of the human stare—on humans.

In his book, *Alfie Darling*, Bill Naughton remarked about the commonness of eye evasion by most people when driving a car in a crowded place: "... the secret of driving in London is never to catch another driver's eye. Whatever you do, never look at him when one of you has to give way ... nobody can get at you in any way ..." A large number of people are apparently affected by the human stare as an act of aggression and take pains to avoid eye contact.

People aren't the only animals to utilize the power of the stare: cobras are said to hypnotize their prey by staring into the victim's eyes. My dog stares at other animals she considers fair game; the lion stares at me—quite probably for the same reason.

Yet again, my house cat must be exempted, along with her domesticated relatives. If I stare at her intently, she will return my stare for awhile, then slowly look away. She simply isn't intimidated by my powerful human stare. Yet if another cat or dog stares at her, she takes it very seriously, so I'm quite sure she recognizes a stare as aggressive. (I cannot know why she is not intimidated by my stare, but I've always surmised that it's simply her conviction of relative equality between us.)

As an act of aggression, the effect of the stare can be reversed as well. In attempting contact with a shy animal, if I deliberately lower my eyelids a bit and avoid direct glances, I've found that such animals will come quite close. I learned this from a nervous cat named Bruce. He became so enchanted with my half-closed eyes, he climbed into my lap for a closer look. The position of my eyelids apparently declared a denial of aggressive intent. We became fast friends from then onward. (Conversely, households where the family dog is stared at often creates quite the opposite situation; the dog never quite trusts his own people.)

The most common human associates in the animal kingdom—the cat, the dog, the cow, the horse and so on—are socialized species. And in our associations with other beings, we subconsciously learn to watch their eyes and ears for signs of irritation, annoyance, and anger. Whether we are conscious of them or not, such signs are usually clear across the boundaries of speciation—but *only in the social animals.*

Such signs don't exist in a good many animals: a rooster's face wears exactly the same expression when his head is on the chopping block as it does when he is crowing in the morning; snakes wear the same expression on their faces in hasty retreat that they do an instant before attacking. Everyone seems to tacitly understand this.

But this is as true of bears. Every year, at least a few people are badly injured or even killed by bears because too few of us recognize this difference, which I finally learned is a difference in facial muscle development. The bear seems unintentionally treacherous to us because we've come to expect standard warning signs before an attack from another animal: ears laid back, change of eye expression, perhaps a furrowed brow or a raised lip—some sign that the animal is about to attack.

With the bear's jolly dog-like appearance, we tend to expect similar visual warnings, much as a dog would give us. But the bear cannot change his facial expression, any more than the rooster can. The facial musculature to change expression simply isn't there.

Being highly social animals, people have a lot of facial muscles, more than most animals. However, this is a difference of *degree,* not of *kind,* with the exception of a few small details.

We cannot lay our ears back as so many other animals can. In turn, most other animals cannot frown. As far as Darwin was able to discover, the rest of the primates don't make much use of the corrugator muscles in the forehead.

Mark Twain deserves the credit for pointing out yet another small difference in people: "Man is the only animal who blushes—or needs to."

After recovering from the humor of that remark, I recognized the sharp insight into a visual signal of humanity that (so far) seems to be exclusively human: blushing. While a blush is visible in every race of human, the tendency to blush diminishes directly with the aging process. A youth is far more likely to blush than an older person.

There is, however, an analogous behavior in the canines that serves the same function as human blushing. When a puppy commits a social error, he will roll onto his back, whimper and urinate a few drops. This puts the older dog on notice that the offender was only a puppy—a mere infant. Large size in the offending dog has nothing to do with this reaction—age is the only consideration. As Dr. Lorenz points out, "A bad-tempered fox terrier treats a young St. Bernard as a helpless baby even if it is twice his size..." The older dog is put on notice that the offending dog was just a pup— and the offense is taken less seriously. The success of the transaction is guaranteed by the puppy's behavior. (Why some older dogs behave this way has more to do with the domestication process, discussed in another chapter.)

The puppy's standard behavior and the young human's quick blush serve identical social functions in the same sort of situation: to put the older being on notice that the offender is young, that the social crime should be weighed in light of inexperience.

As a genuine social difference between people and other animals, blushing doesn't really seem very important—but it currently appears to be a genuine difference. It serves humanity well in its function as a visual signal, a request for a merciful judgment direct-ed toward the observer.

People are continuously alert for such social signals, both between each other and in any association with the rest of the ani-mals. A smile of greeting, a frown of disappointment, the simple raising of an eyebrow, all have special meanings to people. And peo-ple put a great deal of effort into learning to read expression in another's eyes, whether it be the smile at the postman or the raised lip of a neighbor's dog. We all watch carefully for such visual signals. And our skills are particularly focused on eye expression: "...his eyes were glassy/dark/wide with confusion/anger/fear."

Such signs are usually readable from one person to the next. And because they're so easily interpreted, I truly fail to see why all such similarities between human eye expression and the expression in any other animal's eyes should be automatically dismissed on the grounds of anthropomorphism (ascribing human motives into another ani-mal's behavior). If a person's eyes change predictably in response to specific emotions, it seems more likely than not that all other social animals' eyes alter expression in similar ways to similar stress. In our ongoing efforts to extract humanity from the animal kingdom, we may be denying ourselves a genuine tool in understanding other beings. And our denial of the analogy may be just another lofty way to declare people as separate—and it may not be real.

There's one last point about eyes. When I look at a photo of a seal pup, I am moved by his moist, round, expressive eyes. His eyes appear to me as I believe my own eyes would appear to another per-son, were I pleading for that person's support. Expression in eyes is very important to me, as it is to most people. I would be far less emotionally moved by a photo of a bird or a lizard. I want the feed-back eyes can provide, even though I feel strongly that the bird and the lizard are of equal importance to the seal pup—even if their eyes don't tell me so. I like best the feedback provided me by eyes that appear similar to mine. When it comes to providing support for one species or another, my conviction about the importance of the bird and the lizard (or the snake and the alligator, for that matter) is intellectual—but the eyes of the seal pup attract me emotionally as well as intellectually. His eyes matter to me.

The seal pup's eyes can move me emotionally because they look so like my own eyes. His eyes may suffer a different sort of color distortion than mine; they may see more clearly in darkness than mine; his eyes may or may not have similar lenses to my own. It doesn't really matter. His eyes look enough like my eyes to convince me that he perceives things pretty much as I do.

In the area of optical illusions, it seems that people (myself included) have differences from the other animals—but those differences are few—and small.

People have larger genuine differences, and one of those is the subject of the next chapter.

CHAPTER 5
CENTAURS, SATYRS, & SEX

*B*y this time I'd come to realize that none of our differences would turn up where I thought they might be. In the province of reproduction, we people do have some dramatic differences from most other animals; in one instance, we are unique in the whole animal kingdom.

Mythical history is liberally sprinkled with tales of centaurs (a cross between a man and a horse), and satyrs (a cross between a man and a goat—or a sheep). With my thoroughly modern arrogance, I automatically dismissed these tales as preposterous—at first.

Still, no matter where I looked, they kept cropping up. The stories came from more than one continent and extended back thousands of years. I kept dismissing these tales, believing them to originate from the, shall we say, indelicate personal habits of our ancient ancestors. Genetic laws seemed too firm for such beings to have ever existed.

What I learned was that some of the genetic laws aren't quite as firm as I'd believed. And I began to wonder all over again about the centaurs and satyrs from our dark past.

I'd always believed that crosses between one species and another simply weren't possible. It was a surprise to find that experiments have proven such crosses are quite possible. As example, the University of Utah, through their College of Agriculture Department, successfully crossed a female goat (doe) and a male sheep (ram). The pregnant doe not only gave birth to live young; the resulting young (who looked like a slick goat) grew to adulthood and gave birth to a set of twins.

I learned there'd been similar experimental crosses through the University of Nebraska, but that the resulting animal was infertile.

Some *combination* animals are less surprising. Wolves, dogs, and coyotes look so much alike, no one's very surprised to find out they can breed in any combination, resulting in fertile offspring. But more than forty years ago, when the world-renowned ethologist Konrad Lorenz, suggested that the golden jackal played a heavy genetic role in many of our modern dog breeds—that surprised quite a few people.

While the jackal is a definite member of the Canid family, he doesn't look too much like the other three members mentioned above. Eventually, Dr. Lorenz modified his original theory in favor of strictly wolf ancestry for modern dogs. And the firmest evidence that encouraged him to alter his theory was the extra set of chromosomes in the golden jackal: wolves, dogs, and coyotes all have 78 pairs, while the jackal has 80.

Dogs are unquestionably a creation of people. Their original conception was accomplished by modifying existing wild animals to suit human needs. To a very real extent, these modifications continue today, though the fundamental principles practiced have been the same for thousands of years.

People have created other animals over time. The mule is as much a human creation as is the dog. The mule is a specific cross between a male donkey (jack) and a female horse (mare).[5]

Mules are still highly prized around the world: they climb hills better, are less inclined to illness and injury and, finally, they are far less excitable than horses. The mule is a combination animal of long-standing. Records of mules go back to 1500 B.C.

The mule isn't an animal that would happen without some human intervention. But such intervention may be as mild as keeping horses and donkeys in the same pasture.

Still, there is the problem of chromosomal differences: the donkey has only 62 pairs of chromosomes and the horse has 64. The resulting mule has 63 chromosomes, which should make for automatic infertility in the mule. Nevertheless, dating as far back as 350 B.C., there have been scattered reports of fertile mules. The ancient proverb about mules, that they have "no pride of ancestry and no hope of posterity" has long been open to question.

In the 1920s, Texas A&M University acquired a female (molly) mule named Old Beck who reproduced twice in her lifetime. When bred to a stallion, Old Beck's progeny was said to look like a horse; when bred to a jack, her progeny was said to look like a mule. There were a variety of scattered reports of mules foaling in the intervening years, but generally not too much that was new.

On July 6th, 1984, a molly mule named Krause, owned by Mr. and Mrs. Silvester of Champion, Nebraska, gave birth. Testing has proven that Krause is indeed a mule—not simply an odd-looking donkey. (Her dam was a Welsh pony mare, her sire a jack.) When she gave birth to a live foal, she startled everyone and her foaling (and the resulting foal) received national attention.

Aptly named Blue Moon, her foal tests chromosomally as a mule. In the scientific tradition of naming hybrids, the correct title for Blue Moon is that he is a *donkule*: male animal species' name by female animal species' name.

The San Diego Zoological Society has been leading the effort to answer some of the questions raised by this incident. Meanwhile, Krause has not been deterred by all the fuss. She gave birth again on November 11th, 1987. Her latest foal, White Lightning, is also undergoing chromosomal testing.

With all the fuss about chromosomes, and specifically the difference in number between the horse and the donkey, I find myself wondering again about Dr. Lorenz's theory on the golden jackal as a part of the ancestry of some modern dogs. The chromosomal difference between the donkey and the horse appears no greater than the difference between the dog and the jackal.

By no means are these the only examples of crosses between differing species. On the Russian steppes, the Przewalski's horse (pronounced Pshe-vall-skee) has long been interbred with feral (domestic-gone-wild) horses. While both animals are horses, the Przewalski's horse has 66 chromosomes, whereas the feral horse only has 64.

Captive zebras have been successfully crossed with horses many times. And some of the zebras have dramatically different chromosome numbers from the horses: the Cape zebra, as example, has only 32 pairs of chromosomes—half that of the domestic horse.

Some animals, most notably those that have been domesticated for generations, will voluntarily cross the boundary of speciation for breeding. Yet most of these animals (who will cross that boundary) at least look quite like one another—none so dramatically different as a man and a horse (centaur) or a man and a goat (satyr).

This look-alike factor gets on thin ice in some areas. Lions and tigers look alike only in the most general way: both are huge felines. But they have been successfully crossed. The resulting animal, called a *liger*, isn't even considered remarkable anymore. In recent years, a fertile liger gave birth in Frederick, Maryland, to a litter of three *tilis*. They were sired by a Bengal tiger.

With so many unusual crosses, it takes an effort to remember that zoo directors have great difficulty getting animals of the same species to breed in captivity. By far, the majority of zoo animals don't want anything to do with breeding under captive conditions. It's little wonder that so few animals cross the line of speciation for breeding; it just doesn't occur, as a rule, even in more natural situations. Part of both of these problems can be explained by xenophobia—rejection of strangers.

Some of the reasons for xenophobia as a breeding problem, especially in the crossing of species, may be a simple failure between the animals involved to recognize each other's formal breeding protocol. Among some species, these are quite exacting. As only one example, a male penguin rolls a stone to the feet of his intended. If she accepts the stone, a pairing has been established. But if she rejects his offered stone, he takes it to someone else. (Engagement rings come to mind here.) Recognition of each other's gestures encourages pair bonding between two animals. At the same time, it *dis*courages mispairings between animals that are similar but not quite the same.

A fundamental principle of the process of domestication is to dampen these gestures; in fact, a willingness to allow breeding manipulation is a vital key to the process. The result is clearly demonstrated by the incredible variety of dogs available around the world, running the gamut in size from the chihuahua to the mastiff, from the red setter to the black Newfoundland in color, from the dachshund to the poodle in shape.

Since humanity is the manipulator of the breeding of many other animals, it's less surprising that people, too, will cross the line of speciation for sexual gratification. In conditions of isolation—be that isolation physical or psychological—people will cross the line of speciation. Zoophilia (the sexual involvement of people with other animals) isn't all that unusual.

And it's zoophilia we're most inclined to believe was the origin of those cave wall paintings of centaurs and satyrs in the ancient past. If you are satisfied that such behavior was common among our most ancient ancestors, then the puzzle is solved for you.

But it's a quick-fix answer. By the time I got this far into this subject, I began to challenge that easy explanation. There were too many holes in it logically. The tales of combination animals and the cave paintings of them extend into modern times from *well before* the domestication process began. Wild horses, wild goats, or wild sheep were not likely to let people (or any other predator) get close enough to touch them under any circumstances; not while they were alive anyway. And that bothered me. Then there's the seemingly small detail that the earliest domestic animal, the dog, was never involved in any of the combination animals. While the cave drawings of combination animals/people have been around for a very long time, those times were long before the domestication of horses, goats, and sheep. So whyever was the dog missed?

It seemed an insoluble enigma to me until I read *The Golden Bough* by James G. Frazer. The practice of dressing up in animal hides for ceremonies is at least as old as the art of combination animals found in caves. And it finally dawned that the early custom of dressing up in animal hides for ceremonies might be the explanation of those drawings as opposed to zoophilia.

Putting myself in the position of that ancient group's artist, I can well imagine that if I was going to spend all that time to paint a picture for posterity, I was going to make it quite clear that my chief really did look like Pan (a satyr) on that big feast day. I know what goat feet look like (I've eaten enough goats to know) and the chief would be flattered if I made it look like he really was the deity on that day. (Please notice this all applies as well to the centaur—a combination horse/man.)

Once I understood this possible explanation, I suspected the dog was left out of these combination animals because zoophilia didn't occur to anyone until more recent times. Early people were hardly stupid, or we certainly wouldn't have come as far as we have. And they couldn't possibly have gotten close to another animal—certainly not close enough for breeding—unless the other animal was dead. It doesn't take a genius to perceive that dead animals don't give birth.

I think the dog was not in any of these ancient combination animal drawings because zoophilia didn't occur until much more recent times, when more animals were domesticated. Without doubt, zoophilia began in ancestral times, but I'll bet it began in far more recent times than we modern people have been inclined to believe.

Zoophilia is, after all, just a specialized word for crossing the boundary of speciation for sexual gratification, but used only for humans crossing that line with any other animal. Since people are not the only animals that cross that line, such crossings cannot be considered an exclusive difference between people and other animals. But people do have other sexual practices that I thought seemed odd among the animals until I looked quite carefully through the rest of the kingdom.

There are unquestionably people who believe homosexual pairings are unique to humanity—but they are not. According to Vitus B. Dröscher in his book, *They Love & Kill*, "...homosexuality is a common phenomenon in nature." In *The Birdwatcher's Companion*, Christopher Leahy mentions several times that there isn't anything remarkable about same-sex pairing in birds, though it most often occurs when there's a skewed sex ratio in a given population. A long-term study by the Department of Agriculture of the United States found that some domestic rams will not copulate with ewes, though they readily breed with other rams. Homosexuality is by no means unique to people.

Rape, however, is truly uncommon elsewhere in the animal kingdom. Try as I might, I could only find one example of another animal who raped regularly: there's a mosquito in New Zealand whose normal breeding practices involve what people would call rape. (The male copulates with the female while she is still in the pupal shell.) In the majority of species, females *must* cooperate in order for copulation to take place. Psychiatrists have long been telling us that rape is an act of violence anyway, having very little to do with sex. (It's just physical abuse with a different *tool*.)

Fortunately, that mosquito saves us from owning rape as an exclusively human trait. Although there is a good deal of homosexual rape among other animals, that's not quite the same thing. It's a common ending to dominance battles, notably among mountain sheep, some bears, and some antelope. In their way, such rapes serve as solid evidence that psychiatrists are right: rape is first and foremost an act of violence.

Rape has another odd stigma, in that it appears far more common than it actually is. The usual breeding scenario among other animals (and especially the mammals) is of a female fleeing an aggressive male; she seems quite determined to outrun him. Her running seems to cause great excitement in the closely following male and, immediately upon catching her, he mounts and ejaculates. Mission accomplished.

It's only watching the same group of animals over time that I began to notice males pay very little attention to females who aren't in estrus (heat). And when a male is just being a social nuisance to a female of his kind—out of breeding season—she outdistances him with considerable ease. The only time he seems speedy is during the breeding season. And even then, he cannot keep up with an uninterested female—she's far too fast. (I have witnessed this phenomenon in sheep, goats, cattle, and varied other domestic species.)

Appearance of rape or not, this speedy breeding routine is a logical practice among most animals. Historically, any animals that took too long to copulate were likely eaten by a predator who caught the copulating pair by surprise. For thousands of generations, the faster copulating pairs survived; the slow ones didn't. Most breedings seem fast by our standards, but humans rarely are forced to breed where a predator is hanging around, waiting to catch us unawares. For most animals, faster breeding practices mean they get to stay alive. And they invariably produce young who carry on the tradition.

Sex is thoroughly tied to myriad subjects that have become inevitably slightly separate over time. One of these is pair-bonding. Grossly generalizing in the definition of terms, monogamy and polygamy are both quite common throughout the animal kingdom: some individuals of some species bond one-to-one for life; others bond only for weeks, days, or moments, then never see each other again; still others bond into groups.

When I look twice at these bondings, I see that about half of the variations fit the word *monogamy* equally as well as the word *polygamy*. So the words themselves become irrelevant. And when I look at the possible variations, I can also see that various people, in various times and places, have practiced all of these possible bondings—and then some.

For millennia, people have argued about whether humanity is fundamentally monogamous or fundamentally polygamous. Examining the sort of pair-bonding most common among civilized people today, I perceive it as a kind of serial monogamy, where there is strict pair-bonding with one mate, then a long or short period of loneliness, followed by another pair-bonding. There's a ton of evidence to support strict, lifelong monogamy as being the norm in people, and another ton of equally solid evidence to historically support polygamy. Yet I suspect that some form of *serial monogamy* has been most common at any given time throughout human history.

There are often clues to normal bonding behavior in any animal's common reproductive behavior—through management of the young. Human young remain in a state of dependency for an extraordinarily long time—far longer than most mammals. Anywhere in the animal kingdom, when both parents care for the young, those parents stay together at least until their young have been successfully raised to physical independence. A large degree of physical independence is reached in human children by age three.

By physical independence, I mean that many young animals (including humans) are no longer solely dependent on their dams for milk; they are capable of walking on their own; if pressed by circumstance, youngsters are at least capable of recognizing and locating food. And they are quite capable of putting it into their own mouths. While it's a low level of independence, it grows right along with the child, reaching its peak at puberty.

A quick example of the resulting maturity achieved by puberty would be the formation of youth gangs, in which children manage their own lives without any adult supervision at all. Similar events are a constant development at this age: membership (as an individual—not as part of a family) in clubs, in schools, and in social groups bearing no ties to the biological family.

This takes place in nearly every culture in the world, though cultures influenced by European traditions enforce child dependency far longer than other cultures insist on. (In those cultures not influenced by Europe, youngsters are married and managing their own families by puberty.) In industrialized nations, puberty more often represents the beginning of high school, perhaps a long engagement or a college education. This enforced dependence, past the initial stage of breeding age isn't even unique to us: various animals practice it, where one generation helps to raise the next generation to puberty before that first generation is encouraged to leave and start their own families. Beavers come to mind here, as do a variety of birds.

However, people make no distinction between the genders of these young in relation to dependence/independence. And *that's* uncommon.

The rest of the animals who maintain small family groups (as people do) make definite distinctions between the genders of their young, in regard to their dependency. Female wolves, for example, mature sexually a full year before their male littermates. The young female is likely to be mated to an outside male long before her brother even reaches maturity. This reality of differing maturing rates between males and females nicely broadens the genetic pool among wolves. Under normal conditions, young mammals surrender their family ties voluntarily as they reach maturity. Even more than that, the young seem to automatically seek their independence, to break away from the family unit just prior to sexual maturity.

Something like this seems to be common in the human family unit. Our own young, male or female, at puberty often are overwhelmed with a desire for adventure. Traditionally, this is expressed by a desire to join a traveling circus, become a movie star, or start a rock band—anything that might lead far away from home.

European-influenced cultures tend to ignore this natural tendency. But the less urban cultures address this natural inclination by providing hunts or walkabouts[6] or something of the sort, to establish a rite of passage into legitimate adulthood, a more autonomous lifestyle, an independence from the established family unit.

We people tend to ignore our most basic biological tendencies, burying them in cultural tradition in ways that make them hard to discover. But we are by no means the only animals that have problems in the area of reproductive behavior—and that surprised me. I'd always vaguely believed that most of the animals had smooth systems that never failed to function properly, that other animals unerringly proceeded from sex to reproduction to the successful rearing of the young. But one recent study of reproductive behavior tackled an old problem, shedding some new light on the practice of infanticide. Sadly, infanticide isn't at all rare in the animal kingdom.

Since the initial study was done on lions, lions will make a good example here, but keep in mind they are not the only animals who do this. A lion pride (family unit) normally consists of numerous females and one patriarchal male. In his position of dominance, this male may be challenged at any time by an outside male, a nonmember of the pride.

If the patriarchal male loses his battle with the outside lion, his position within the existing pride—his family—is taken over fully by that outsider. The former patriarch is driven away from the group; he not only loses his high status in the pride, he also loses his entire family. He cannot maintain any contact with the group.

The *outside* lion has taken the former patriarch's social position, his mates *and* his progeny. In his new seat of power, the usurping male may simply kill the cubs within the pride. While the females certainly object to his behavior and try to protect their young, the bottom line is that the usurper often succeeds in killing some or all of the cubs that were in the pride when he took over.

Until recently, those who studied lion prides (and other species that practice such takeovers from one male to another) always considered these killings an aberration. To people in general, the practice seemed just plain cruel; to scientists, it seemed destructive to the species as a whole. But since this aberration cropped up with monotonous regularity, it was formally reexamined. And eventually, it was recognized that such seemingly destructive behavior might actually make evolutionary sense, in the long run.

Looking at this same situation in a new light, when the *new* male lion takes over an already established pride, he does so by defeating the last pride leader in battle. And as long as the cubs of that former pride leader are present, still nursing off their mothers, the females don't come into estrus (heat).

But if those nursing young die, the females quickly revert to the breeding phase: they come into estrus much sooner than they would have if the cubs were still present. And the cubs following breeding will belong to the new male.

I do not mean to imply that the new male does a lot of thinking about this. On the contrary: his bullying, abusive behavior is his instinctive personal effort to show off his new power to his female subordinates. But the end result is the same: his genes become a part of the pride's gene pool sooner.

Although this practice seems unduly cruel, if the former pride leader lost his patriarchal position due to some flaw in his genes, his progeny will die without ever passing that flaw to yet another generation. The new sire's ability to win the dominance battle may have a genetic advantage to the group as a whole.

Lions aren't the only animals that do this. The practice seems most common in the carnivores and omnivores who maintain small family groups. (I've wondered from time to time if there's any vague similarity between this and the currently high number of child abuse cases being reported, especially in new family formations, where the children present were created with a former partner.)

Mainly because he is larger, the male lion succeeds in bullying the female lions, possibly killing their cubs in the process. Sexual dimorphism (two types in one species; in this case, referring to size difference between males and females) is common throughout the animal kingdom. Males tend to be larger among mammals, though this is not always the case.

My lifelong assumption about this size difference nearly tripped me up here. I'd always believed that males were larger to protect females and their young. But when I look at the reality more objectively, what I notice is that the male size advantage is used most often as the male lion in the earlier example used it: to battle other males and to control females.

Without question, males of many species use their size advantage to protect females and young from danger. But larger males also use their size advantage most frequently to protect *themselves*.[7]

Sexual dimorphism in size between the genders is common: female jaguars tend to be smaller than male jaguars; female monkeys and female wolves tend to be smaller than their male counterparts. The human difference in size between genders is remarkably consistent in the kingdom of animals.

Once again, human differences from other sentient beings tend to be less numerous than our similarities. There is, however, one absolute unique human difference in this area: humans are the *only* animals with the *potential* to control our own numbers.

Even this issue had its hidden surprise. In every way, the human population explosion seemed to me to be an ancient, insurmountable problem. But the population problem isn't even long established, and it certainly isn't insurmountable. In all the important ways, the problem itself arose last week. Figures vary from one organization to another, but all organizations confirm that the problem is quite new. Using the figures of *Zero Population Growth*:

- Human population didn't even reach its first billion until *1800*, due in part to the massive plagues and the greater instability of human life before that time.
- 130 years later, in *1930*, the number of people had doubled to two billion.
- In *1960*, a mere thirty years later, the population of humanity had reached three billion.
- 15 years later, in *1975*, human population had reached four billion.
- 12 years later, by midsummer of *1987*, we passed five billion.
- Just nine and a half years later, in *1997*, we hit six billion.

The enormous jump between 1800 and 1930 is due, at least in part, to Dr. Jenner's publication of the principle of vaccination. Between 1930 and 1960, world population jumped incredibly: antibiotics came into being, food distribution became mechanized, and various scientific and medical advances added to the stability of human life. Oddly, that time period represents some of the greatest losses of human life as well: an estimated 55 million people lost their lives in World War II alone. But the Baby Boom Generation, born between 1946 and 1964, produced 78 million more people in the United States alone, which certainly filled in the gaps. Because of that jump, increases have fallen into a predictable geometric pattern.

Around the time the human population reached a billion in number, an economist named Thomas Robert Malthus suggested that population increases in a geometric ratio—which is seen in the foregoing numbers. (The time it takes for the human population to expand yet another billion grows continuously shorter.) The numbers also make clear the reality that this problem is of recent origin. Human awareness of it is newer yet.

People aren't the only animals who have had to deal with overpopulation problems, though usually, for most other species, such problems are localized. The most common response in other animals is *emigration*. (Migration is where animals leave an area for a time; emigration is only one way.) Anyone who has lived near a major stream or river for a long time talks about the astonishing day all the deer (or mice or squirrels) swam across the water unexpectedly.

It's emigration that's behind the famous tales of the lemmings of Norway, casting themselves into the sea. What's really going on is that the lemmings are moving away from heavily crowded areas, moving away in all directions, crossing fjords and mountains as they go. Some of these emigres make it as far as the sea and, apparently in the mistaken assumption that it's merely another fjord, the lemmings dive right in and swim away from shore.

Humanity once had that option, and used it, to emigrate to the four corners of the earth. Perhaps people can eventually emigrate into space, but the time available where such a solution might actually benefit us is shrinking. And frankly, I don't believe people will go through this galaxy or another, trashing planet after planet, until forced to return to the original problem: human overpopulation.

Some animals deal with overpopulation quite differently than through emigration. The lynx and the snowshoe hare are frequently used to explain problems of population dynamics: with cyclic regularity, the number of hares in a given area increases. (The reasons behind the increases aren't yet understood.) According to naturalist and filmmaker Marty Stouffer, fluctuations in the snowshoe hare population may vary as much as eight hares to the acre on the high side. Those numbers may drop as low as one hare per twelve acres when on the low side.

The lynx population follows this cycle closely, about one year behind the hare. And that means that when the number of lynx rises, there is a serious period of starvation and death until the pendulum swings back. No matter what else is going on, the lynx can never afford to become as numerous as the snowshoe hare. The entire cycle for these two species averages completion in nine to ten years.

People have no similar advantage. We don't have a lynx or a hare to help us control our numbers.

But we do have ourselves. People are quite good at solving their own problems, and only recently a possible solution to this population difficulty came about via birth control. Our current and real problem, in most ways more important than anything else, is *distribution* of existing birth control methods.

We need to buy time. At the moment, we're still at the early finger-pointing stage, inevitable with all our problems. We point everywhere else (usually over the heads of our children) at other places as the source of the problem. Advanced cultures point accusingly at less advanced cultures without offering any assistance or practicing much birth control themselves. The continent of Africa points at India which, in turn, points at China, a nation that has begun serious efforts to control its population.[8]

And all the while, the overpopulation problem grows. If you celebrated your sixty-third birthday in 1993, our global population has grown about 3.6 billion people in your lifetime. If you were born in 1960, world population rose by 87 percent by your thirty-third birthday. And if you are quite young, having been born in 1975, you commemorated your eighteenth year in a world with 40 percent more people.

Dr. Malthus' paper on population problems became known as the *Malthusian doctrine*. He had put forth the idea that crime, war, and disease are directly related to population increases. In other words, *all human problems, without exception, relate to our recent population explosion.*

While the solution may be personal, the problem remains universal. And it follows that, if we can solve this biggest problem, we'll be taking a giant step toward solving many of our other problems.

Considering this reality, remarkably little study has been done on the behavioral changes among animals in reaction to the effects of overpopulation. (The biological changes, as opposed to the behavioral changes, has been the general focus of most studies.)

Among people, behavioral changes have been subtle, but discernable. We value each other less and less, as time goes on. If the Malthusian doctrine is correct, and crime is a direct response to population increases, it follows that defense of property is an extension-reaction to crime. Surprisingly, loneliness seems to be a general reaction to overpopulation: it becomes more difficult to relate to one another as crowding increases.

The enormity of the problem of overpopulation is only realized after we've had our own children. I know of one public figure in the United States who has eight children, and is now preaching population control. Because people don't recognize the dramatic problems of overpopulation *before* they have their own children, they often have a shoulder-shrugging reaction to current figures; the problem is then perceived as something they can do nothing about. (Such people could at least contribute financially to make birth control available to those who want it anywhere.)

Usually, when overpopulation of the human animal is mentioned, the most frequent concerns are shortages of food, water. Without question, these are vital concerns that will have a major impact on human survival, in both day-to-day and long-term existence. These specifics will effect the quality of life—perhaps even the possibility of life—for all living beings, most especially those of the animal kingdom.

But Malthus also mentioned disease in direct relation to population increases. Little was understood about disease in Malthus' time but even then it was expected that disease would be an inevitable controlling factor in human population, as it is among all animals.

Although many biologists write about the population problem, the majority flirts with the subject in a way that offends the least number of individuals while successfully criticizing the highest number of people. In more recent years, however, disease has been mentioned more frequently as an inevitable natural factor in population control.

When any animal population becomes dense, their numbers are usually reduced by disease. For example, when Smith Island (off the coast of Washington) became overrun with rabbits, those rabbits became undernourished, then diseased, directly because of their population density.

A slight variation of the same theme, again involving rabbits: in 1859, Thomas Austin had a dozen pairs of European rabbits shipped to his Australian home. In less than forty years, that little shipment became the focus of the entire nation of Australia. Something similar happened in New Zealand where a single ranch went from twenty thousand sheep to two thousand sheep in only four years—directly because of a rabbit population explosion. (The rabbits ate everything green and growing, leaving nothing for the sheep to eat.)

When *myxamatosis cuniculi*, a rabbit virus, was discovered in Brazil, news of its high fatality rate was rushed to those two rabbit-beleaguered nations. And it represents humanity's first efforts at biological warfare in deadly earnest. The spreading of the disease was quite successful in Australia. New Zealand had less success, and had to work out other methods to control its rabbit population.

Such biological controls would not have been successful if the rabbit population had been sparse. The Australian rabbit extermination was successful only because the rabbits were so numerous; the disease spread readily from one rabbit to the next. The same technique was less successful in New Zealand because that nation lacked the necessary insect population to easily transmit the virus from infected individuals to healthy ones.

Why disease automatically becomes a factor in reaction to population increases is not well established yet. Many biologists believe it has to do with the mutation of disease organisms. For example, as the number of rabbits increase, the number of bacteria *in* the rabbits also increases. When bacteria rise in number, mutation becomes more likely as in any organism. The result is disease.

In any kind of animal, disease only becomes a problem when a given population reaches a critical number. That critical number appears to be different in each kind of animal. The number is relative to other factors as well, such as habitat: one hundred million[9] bears in the state of Alaska would be a clear overpopulation problem, even in that large an area; one hundred million mice in the same area might not even be noticed.

In our size category, we are by far the most numerous animal on the planet. We are currently able to maintain inflated human numbers through technological advances: productive food systems, a new awareness of the importance of global environment, medical and scientific development, new management techniques of dealing with this and that. But all of these—*all of these*—are only stop-gap measures, postponing dealing with the basic problem that creates the need for *all* of them. The world and the human overpopulation menace will dissipate when birth control becomes widely available and popular everywhere.

Until then, the realities of the potential loss of life are horrifying. Unfortunately, disease isn't the least selective in its choice of victims and human losses will undoubtedly be great and devastating.

But not *necessary*. People are absolutely separate from the rest of the animal kingdom in this way: humanity is facing the most gargantuan problem ever faced by an animal—and it also has the ability to deal with it. Human awareness of the problem is still too new to assume it's insurmountable. But it is a problem equal only to human problem-solving abilities.

The overpopulation problem—and its solution—represent *the*

first major difference between people and other animals. People do have other similarities and differences with the rest of the sentient beings. But this is an important difference: it's the first difference in *kind*, not simply in *degree*.

Part II: Behavior

CHAPTER 6
DOMESTICITY

W hile the problem of human overpopulation is quite new, it was only able to occur because of the ancient concept of domestication. (Another example of human dependence on the rest of our kingdom.) Had people never domesticated some other animals, human life patterns, technological advances, even our progress as a species might never have occurred; at the very least, these things would be vastly different from what they are. The results of the domestication process, begun so long ago, can hardly be exaggerated.

Our ancient ancestors could get a better night's sleep, resting in the assurance that the household (or cave-hold) dog would sound an alarm warning of possible invaders. The truly incredible notion of riding an alien being (such as a camel, a horse, or an elephant) from one place to another certainly helped humanity spread quickly to the farthest reaches of the globe. And most of all, the domestication of animals made human urban centers possible: in one way or another, cities—large concentrations of people—would not be possible without other animals. Either those other animals are used directly as base sources of protein (in the form of milk, meat, or eggs) or they are utilized in the production of food (in the form of fertilizer-producers or work animals). While some cities use fewer animals than others, all urban centers use some.

Though the process of domestication has been of such enormous significance, we haven't been able to learn much about how it began. Why did people in five isolated regions (northern Europe, Indochina, Africa, southwestern Asia, and South America) all suddenly begin confining animals?

Even the types of animals confined varied greatly: people of northern Europe confined reindeer; people of Indochina confined a large variety of animals, including the water buffalo; people of Africa domesticated the modern house cat; people of southwestern Asia kept a wide variety of animals in captivity, including cattle and horses; people in South America domesticated animals such as the llama and the turkey.

It appears that domestication was simply an idea whose time had come. People, scattered as they were, may have all reached the same plateau of development at about the same time.

As to when the concept began, the theoretical era has been pushed steadily backward in time: not long ago, the accepted figure was about four thousand years ago; shortly after, the figure was pushed back to six thousand years. Currently, it's assumed to have been about ten thousand years ago.[10]

No matter where it all started, it's still going on. The laboratory rat is a relatively new creation since modern technology created the need for such an animal. Some families of African lions (*Felis leo*) have been bred in captivity for enough generations to be considered near true domestication. And Africa has many sites of activity right now where ongoing attempts to domesticate native animals are being made to replace cattle in food production.

In the African landscape, cattle cause all kinds of serious problems: they overgraze native plants; they create erosion and spread desert wherever they go. Cattle need to parade back and forth to watering holes, and they destroy quite a lot of vegetation in the process. Conversely, native African animals evolved with local vegetation, and their grazing patterns tend to increase plant growth. Quite a few of them get the water they need directly from the plants they eat so some never seek waterholes. Because native African animals evolved with the environment, they cause minimal damage to it. Current efforts to domesticate native animals for human use will have large benefits.

Described simply, the domestication process is accomplished by teaching some animals (of a given species) to live and breed in close proximity, both to each other and to people. Phrased in this simple way, domestication seems to be wholly a behavioral change. It is not. The entire process, and its result, is much greater than is easily apparent. And it always involves much, much more than the supposedly simple task of taming a few wild animals.

Ultimately, the process of domestication is easiest to discuss when using the family dog as an example: the dog is both domestic and tame. The farmer's bull, however, is equally domestic but rarely tame.

Basically, the long-term process of domestication is used to enhance plant or animal traits that people consider desirable such as bigger corn kernels, greater milk production, frequent egg production, and greater tractability in working animals. There are hundreds of other examples. It's also important to remember that many animals have cooperated in the process of their own domestication. A great many others simply won't do that; they aren't in the least inclined toward cooperation.

The problems of getting wild animals to breed in captivity can be extraordinary. This is as true for zoo directors as it is for those trying to domesticate a species. The reasons for the uncooperative attitudes of so many animals are often subtle and frequently undetectable. Sometimes a female simply will not ovulate in captivity;

other times, a female will be in estrus (heat) and the male simply isn't interested. People in charge of such situations often become exasperated and willing to try just about anything. Occasionally, extremes succeed.

Awhile back, there was a media circus surrounding the estrus cycle of Ling-Ling, the female panda that the United States received as a gift from China. For several years, her mate refused to cooperate. Those in charge finally gave up on him and resorted to artificial insemination. And a few years before that, there was a big fuss about Frazier, a male lion that was adamantly pursued by every female lion he encountered, including a number of females who had refused to cooperate with many other male lions. Without a doubt, the majority of such problems occur with wild animals in captivity.

A breeding curve occasionally occurs in a species of long domestication. Not long ago, there was a quiet fuss about a race-horse stallion that refused to breed with the mares presented to him. Because many people had a great deal of money invested in him, they sweated earnestly for awhile. Someone eventually discovered that the stallion had no objection to breeding; he simply didn't want to breed bay mares. (He was a bay himself: overall brown with black mane, tail and leg markings, the color of the vast majority of race-horses.) His managers learned through trial and error that if they presented him with a sorrel mare first, he would then breed any mare brought before him. His preference made for tricky and awkward problems, but those involved were pleased that they'd been able to solve the problem.

Resistance to breeding manipulation is no surprise to many people, unless the animals involved are domestic. A few species have long resisted such efforts after centuries of being considered domestic. The house cat (*Felis domesticus*) is still noted as a resister. Many elephants also resist such breeding though there have been sufficient matings to qualify both species as having domestic strains.

In researching this subject, I was surprised to learn of some other sorts of domestic animals that are less well-known. There is an otter in India trained to catch fish for people. Some of the better-known domestic animals have odd uses: dogs and some pigs are specially trained in Europe to seek out truffles by scent. (Truffles are an underground fungus highly prized by food connoisseurs.) In the United States, some dogs are specially trained to locate illegal drugs being smuggled into the country.

But the oddest incidence of domestication must be the weasel. In early works on domestication, I stumbled across repeated references to the weasel as a domestic animal. Old texts casually refer to the domestic weasel; new books refer to the weasel as a former domestic animal. While it's understandable that early domestications might have gone unrecorded (because so few people could read and write), the domestication of the weasel *was* recorded. This, for lack of a better word, *un*-domestication of the weasel is unique in every way, yet it is barely mentioned in a wide variety of texts.

Weasels were domesticated early in human history as people began to store large amounts of food for later use. Rodents raided and contaminated vast amounts of food. Early farmers looked for animals that would kill off the invading rodents. (Because modern farmers have now learned a great deal about poisons, using animals to protect stored foods is once again known to be a good idea.)

Weasels kill (and therefore control) rodents. Apparently, the early relationship between people and weasels went well; after all, the weasels slowly became considered domestic.

But the weasel is quite capable of taking care of himself. He's one tough little killer, known to successfully attack and kill animals more than a hundred times his size. I suspect some rather nasty incidents with the domestic weasels hurried along the earliest attempts to domesticate the more manageable cat, who is still employed as a rodent control all around the world.

Something like this may have occurred long ago with the dingo of Australia. Although today the dingo is considered a true wild dog, many authorities agree that the dingo and Australoid man (the Australian aborigine)—the only large non-marsupials native on the Australian continent—likely arrived in Australia together thousands of years before recorded history. And with thousands of years having passed, the dingo's status as a true wild animal is understandable; but the weasel only fell off the domestic list sometime in the last century. Weasel status as a true wild animal is puzzling. Weasels aren't even classified as feral.

Feral animals have their own unique status. There have long been official reports (and unofficial rumors) of feral camels in the southwestern United States. The earliest imported animals (1701) didn't survive. They were landed in Virginia and were quite unable to survive in that humid climate. Later, just before the Civil War, the United States' Congressional Appropriations Committee ordered the War Department to spend $30,000 on the purchase and importation of camels, "...to be employed for military purposes."

This second group of dromedary (one-humped) camels arrived in Texas in 1856. These thirty-four camels did so well in Texas that yet another group of forty-four animals was brought in during the following year. All were used to survey a wagon road through the southwest and to carry freight in California.

Being so close to the start of the Civil War, the idea of using camels for military purposes never had a chance to develop. There was too little time to train anyone to work well with the camels. Once the war began, many camels escaped. Others were sold to circuses. Still other camels were employed in mining work and some of them escaped too.

Because the climate of the southwestern United States is so well suited to camels, the escapees did very well there. The last official sighting was reported in 1941 in Arizona. There may still be some feral camels in that region, but regular sightings haven't been reported for a long time.

Another feral animal who has caused far more problems[11] than the camel in the United States is the wild boar. The original fifty boars that were brought into the country in 1893 were genuinely wild animals. Though that original group was shipped from Europe, later groups were shipped from the Soviet Union and Asia. All of these readily interbreed with each other and with feral domestic hogs as well. At least many thousands are well established across the continent; that is likely too modest an estimate. The Audubon Society reports that some pure strains of the original wild groups can still be found.

Earlier, I mentioned that genetic manipulation is a large part of the domestication process. This shows clearly in the difference between wild boars and feral hogs for the first generation (or so). The wild boar has a *straight* tail; the feral hog has a *curly* tail. There are other differences, of course, but this one is a clear example of another standard feature of domesticity: *neoteny*.

Neoteny is a high-priced word meaning the retention of juvenile (sometimes infantile) characteristics throughout adulthood. And neoteny is one of the standard signs of domesticity in any animal. Even the wild hogs have a curly tail when young; as they become adults, their tails straighten out as part of the maturation process. Domestic hogs retain this juvenile trait throughout adulthood; their tails don't straighten as they become adults.

Obvious signs of neoteny show up in dogs (*Canis familiaris*). Wolves (*Canis lupus*) have floppy ears, curly tails, short coats, and light builds—as puppies. As they mature, wolves' ears become erect, their tail carriage drops, their coats and bodies fill out. In comparison, the majority of domestic dogs retain at least some features of wolf-puppyhood throughout their lives. Even when the physical traits of an adult dog look a great deal like those of a wolf, the dog will still retain some of the behavioral traits of a wolf puppy: adult wolves rarely bark, but adult dogs often bark; wolf puppies lick older animals' faces and so do adult dogs. Although the submissive behavior common in dogs is a natural trait of wolves, it's carried a bit further into adulthood in dogs. Even in the instances where dogs do act like wolves, there's usually some variation that clarifies the dog's domesticity.

Another significant physical difference between wolves and domestic dogs: wolf-sized dogs have 20 percent smaller brains than wolves do. It bears repeating here, however, that brain size doesn't seem to mean much; instead, how well the brain is used makes the real difference. Dogs are unquestionably wonderful, intelligent, unbelievably brave animals who share a long and complex history with people as valued friends, guardians, and helpmeets of the highest caliber. Without their long and incredibly versatile assistance, human history would have been quite different. (The wolf, even with his larger brain size, is a far less versatile companion.)

Dogs have been bred for thousands of generations to develop traits that are often only mildly present in their wolf ancestors. As example, a wolf will defend his (or her) own den ferociously; but your dog will defend *your* home with equal ferocity. A wolf can be tamed in a one-on-one relationship, but a dog begins a relationship as a domestic, willing to learn to be friends.

I've talked to people who have raised wolves from puppies; every one of them mentioned consistent problems. Rightly, the laws of all societies restrict the movements of wild animals within those societies. And those laws force the owners of wolves to keep them confined or on leashes at all times. All the people I talked to also mentioned that as their wolves matured they became less submissive, less obedient, a bit grouchy, less anxious to please their owners, and notably inclined toward vigorous independence. As one farmer pointed out, "I can't exactly use him to bring in the cows, now can I?"

Tame wolves highlight another startling facet of the difference between *tame* and *domestic*. Though the process usually goes unnoticed, taming must take place with each generation of domestic animal. Each generation of kittens, puppies, calves, or colts must learn individually that it's safe to be handled by people. The tendency to calmly accept such handling by another species is a genetic trait, but the taming itself remains an individual process. If a puppy does not come into regular contact with people between three weeks and three months of age, he (or she) is extremely difficult to manage or train in the future. Such puppies seem to perceive people as fear-inducing aliens, and it's very much an uphill battle to win them over.

For dogs in particular, the taming procedure has become a substitute for their normal socialization process. (A young wolf learns to look out primarily for his self-interest; a young dog learns to accept that what his people want is of equal or greater importance to what he wants.) Incidentally, the socialization process takes place in the wolf and the dog in the same time frame, between three weeks and three months of age.

In our ongoing efforts to manipulate animals genetically, we have accomplished no less than miraculous results. People took skinny, rangy cows that barely produced enough milk to feed a single calf and, through genetic manipulation, created cows capable of averaging at least forty pounds of milk a day—enough to drown a calf. We've taken birds that normally laid a clutch of eggs in the spring and, through genetic manipulation, encouraged them to lay an egg nearly every day. We've taken the mildly fleet, bulky wild horse and turned him into the sleek, magnificent, swift animal who continuously sets speed records—and then goes on to break them. People have been astonishingly creative at genetic manipulation, often achieving quite unbelievable results.

But not once—not even once—have people been able to produce anything that wasn't already within the animal. We have only been able to build upon what was already there. That's all.

Through the process of domestication, we have genetically

manipulated animals into sometimes overwhelming differences. But in every case, the fundamental material for the changes we made were there to begin with. The sheep had a woolly coat; the otter already caught fish; the cat already caught mice. We have only been able to build upon tendencies that were already present. We have never been able to create a cow that lays eggs or a chicken who milks. The fundamental genetic material to produce wool, lay eggs, run fast, or whatever we want, must exist within the animal before we can manipulate it, before we can exercise our brand of domestication.

Surprisingly, people are not the only animals who keep other animals confined. Some ants maintain herds of aphids, much as we do dairy cattle; the ants herd the aphids to selected pastures, watch over them during the day, and then milk them when they re-collect them at night. Yet another kind of ant maintains caterpillars for a honey-like substance the caterpillar exudes on demand. Both kinds of ants kill off surplus livestock on occasion and continue to tend the rest. Still another kind of ant cultivates crops, much as we raise corn and wheat for future use; these ants raise molds and tend them as diligently as a human farmer tends his crops.

But as far as we now know, none of these other domesticators seem to manipulate their end products genetically. The other species that practice other animal manipulation seem to raise other animals as they find them, leaving them unaltered. Genetic manipulation seems to be our exclusive invention.

With the above in mind, it's less surprising that people have recently recognized that humans as a whole have been changed by the domestication process. In chapter 2, I mentioned the unnerving surprise that Dr. Lilly received when he was testing dolphins for intelligence: he discovered that the dolphins were testing him right back. Recent studies have shown a similar pattern in human development: in the process of domesticating other animals, we've been unconsciously employing similar genetic manipulation on ourselves.

In their book *Between Pets & People*, Doctors Beck and Katcher put forth the credible idea that "Man, not the dog, is man's first domesticated animal." And there's quite a lot of evidence to back their observation. Neoteny—the retention of juvenile traits into adulthood—becomes obvious when we compare ourselves to the rest of the great apes:

Juvenile great apes	*Adult great apes*
short arms	long arms
smooth brow	ridged brow
upright gait	inability to easily stand erect
relative hairlessness	hairy bodies

Besides our physical tendency to resemble juvenile apes, we also have behavioral traits that we consider signs of domesticity in the other animals: one example would be our willingness to breed under almost any circumstance. And while genetic manipulation has been vehemently resisted when any despot in history has tried it, such manipulation has truly always been practiced through social restrictions on pair bonding.[12]

The only certain distinction I've been able to spot between our domestication of ourselves and our domestication of other animals is that we call our own *civilization*. When I look closely at what we call civilization, it's just a stylized variation of domestication.

One of the main elements missing in our domestication process is natural selection. It is no help to the chicken to lay an egg nearly every day; it is no help to cattle in general to have such huge udders, and it is no help to the human race to save any and all genetic deformities for future breeding. But we do all that all the time.

In our own domestication process, we've eliminated natural selection wherever we could. We manage our own breeding programs rather sloppily, I'm happy to say, but we do manage them. We strive hard to encourage folks with birth defects to reproduce; we keep alive the sick and infirm; we try not to let too many people starve to death; we create laws and enforce them to protect the weak from the strong and the strong from the collective weak.

We breed in any season of the year—a sure sign of the success of our own domestication. We breed ourselves selectively, but more for social reasons than for any physical benefit. And because of that and our high numbers, we often achieve spectacular results in the areas of intelligence, beauty, physical fitness, capacities for compassion, artists of all kinds, glorious achievements in vast areas—all unexpected, all unplanned for, all a surprise. While we call the result civilization, it certainly has many of the standard markers of the domestication process.

And we are as deeply dependent on the system itself as are all the other domesticated animals. The twin systems of domestication and civilization are so interdependent that either one would fail without the other.

In *The Lonely Silver Rain*, John D. MacDonald wrote, "In testing any hypothesis, one useful method is to carry it to the ultimate limits of absurdity and find out if it still hangs together." That method would work quite well on the issue of our dependence on our own domestication systems.

As an example, pick a theoretical date to declare worldwide freedom of domestic animals. Every animal confined in a cage, a barn, a pasture, whatever would be set free on this selected date.

The very first thing that would happen is that most of the domestic animals would die quickly. A great many of them would simply have been caught in a climate where they couldn't possibly survive without protective housing. But they have also been genetically engineered to dependency. They are accustomed to functioning on high-quality foods provided by people. Like people, they are also heavily dependent on antibiotics; their immune systems are (at the least!) repressed.

But there would also be other major obstacles: the productive laying hen would quickly waste her stored energies laying eggs; she is unlikely even to be able to feed herself with her genetically shortened and manually bobbed beak. While the milk cow might be able to reabsorb some of her excess milk, the physical trauma of doing so would lower her resistance to any passing disease. The sheep might enjoy his woolly coat in winter, but one season follows another; he would be subject to heat exhaustion and parasite infestation as soon as the weather warmed up.

Humans wouldn't do so well either. Our dependence on domestic animals is enormous. Besides the immediate loss of the high-quality proteins provided by those animals, which would be traumatic enough, we would rapidly lose much of our agricultural lands, currently being kept unforested by "...the ever-munching mouths of the domestic animals."

People would also lose the vast majority of fertilizer worldwide immediately, both through the remains of commercial butchering and the loss of manure. And we would be hard-pressed to keep ourselves fed because of those losses. Everything would change, drastically, and forever without domestic animals. At the very least, it is truly difficult to say who is the master and who is the slave in our current system of domesticity.

The number of people who find the entire practice of maintaining domestic animals distasteful has been growing. Such folks see our current system as embarrassing to civilized humanity. And while there's no logical argument against their ethical position, there is a practical reality that makes the dissolution of domestication forever unlikely. To dismiss domestic animals would hardly be an act of kindness, to *any* of the animals involved.

We can't even dissolve our own form of domestication. We are as heavily dependent on that system as are the other domestic animals.

Taking the idea to its extreme limits again, to demonstrate our own dependence on the system of civilization, I can go back in time (on paper) to Neanderthal people, the last ancestor of ours who exhibited few of our modern domestic features. Neanderthals had the long arms, the brow ridge, and probably the bodily hairiness of the other adult great apes. Even that far back, however, humankind was already beginning to show signs of neoteny: Neanderthals had already developed an upright gait. That feature extends back nearly four million years to when we were much closer to being simple primates than true humans.

The easiest way to make the point would be to switch one modern man with one Neanderthal man. To make their lives easier and to keep the theoretical switch still functional, I'll select each of these theoretical men from a population center of their time: the modern man from a city, the ancient man from a clan or family group.

The first few days of such a switch would be a nightmare of confusion for both men. I don't see local language as an insurmountable problem for either man. Neanderthals had the necessary jaw, throat, and tongue development for crude speech. They had, and apparently used, verbal skills. Because some modern cultures contain small populations of people within their borders who don't speak the national language at all, I don't see any reason to assume that the ancient man would have any trouble making himself understood. Conversely, the modern man could certainly make himself equally understood; the language of fifty thousand years ago was almost certainly simple.

Both men would have some problems in being accepted by the people of such extremely different ages, but again, the Neanderthal would have it a bit easier. In *The Beginnings of Humankind*, Donald Johanson & Maitland Edey state:

> One hears talk about putting [Neanderthal man] in a business suit and turning him loose in the subway. It is true; one could do it and he would never be noticed. He was a little heavier-boned than people of today, more primitive in a few facial features. But he was a man. His brain was as big as a modern man's, but shaped in a slight-

ly different way. Could he make change at the subway booth and recognize a token? He certainly could. He could do many things more
complicated than that.

Neanderthal man in a subway wouldn't cause any sense of alarm
in those he bumped into. He simply wasn't that different.

By contrast, the modern man would seem very different to the
Neanderthal people. During prehistory, it was probably quite
unusual to bump into anyone from outside your immediate family.
In addition the modern man's hairlessness, smooth brow, and short
arms would make him look dramatically different. (I find it odd that
the same situation in both cases would not have identical results.
The reality is that we have a great deal more diversity of physical
types in our times than Neanderthals had in theirs.)

Food would be an immediate problem for both men. The
Neanderthal would be able to see other people eating everywhere,
and he clearly had the mental capability to figure out where the food
might be coming from. Even if he wasn't too quick about figuring it
out, it's likely that some organization would feed him, once his
hunger became apparent.

The modern man's situation would be more difficult. Assuming
he could overcome his immediate problem of being accepted by the
group, he would then need to earn his keep. He would need to learn
fairly quickly which plants were safe to eat; otherwise, he might be
standing in the midst of plenty and never be able to recognize the
edible plants of fifty thousand years ago. In fact, even an agronomist
might be hard-pressed in such a situation; plants, too, have done a
great deal of evolving in the intervening millennia.

No matter how well both men adjusted to this theoretical
switch, I'm inclined to believe both would die fairly quickly. The
necessary adjustments would be too great for either man to manage
more than briefly.

The final killer—in both cases—would probably be disease.
While the Neanderthal man would have a sturdy immune system,

bacteria have also been evolving for the last fifty thousand years. The common cold alone would continuously reduce his natural resistance until it finally killed him; he would inevitably succumb to the virulence of modern bacteria.

While the modern man would be exposed to far less virulent bacteria, his basic immune system has been weakened by modern life. With the low population level of that ancient past, he would be far less likely to encounter very much bacteria from a human source, but even the smallest amount would wear down his diminished natural immunity. Without modern sanitation, modern antiseptics, and antibiotics to help him, he'd be in big trouble fairly quickly. He is dependent on modern standards to supplement his immune system.

Under even mild examination, it becomes clear that the systems we have built up and built upon are as vital to our survival as the domestication system is to the rest of the domestic animals. Our interdependency with them has deep roots in our past.

As human concern about our environment has grown, so has our vicarious understanding of the rest of our kingdom. As we come to understand other animals better, we can see our own reflection in their very existence. Our ability to relate to their needs has made it obvious that their needs and ours are far more alike than different—except maybe in degree, yet again.

Our version of domesticating other beings may very well be unique to us, but it would be splitting hairs to declare it any kind of major difference.

CHAPTER 7
STEREOTYPING

*I*t's been an unending source of amazement to me that many people cannot distinguish one calico cat from another, one brown dog from the next, one bay horse from a distinctly different (to me) bay horse. Such differences have always seemed obvious. It's taken me quite awhile to realize that many people only focus on the similarities between one animal and another. A surprising number of people truly cannot perceive the differences, at least not without spending lots of time and energy noting those differences.

Eventually, I realized that this was the major factor behind stereotyping: looking only at generalities, matching likenesses, and ignoring (or genuinely failing to notice) differences. I've come to think of this way of looking at animals as a prism we look through, denying ourselves the delights of detail, generalizing all that we see.

It also took awhile to discover I had my own blind spots. I've generally ignored birds for much of my life and was astonished to realize that, although I could easily distinguish a starling from a robin, I couldn't distinguish one robin from another. When I spoke to a bird enthusiast about this, he showed me a flock of juncoes (snowbirds) and pointed out the differences from one to the next.

"This one has a greenish hue. And see that one? He has a crooked tail feather. This one here has a distinct way of cocking his head. That one has a white spot on his left wing . . ." He went on for quite awhile. Once I began to see their differences, I had a hard time understanding why I'd thought they all looked alike before.

Then there's fish. I've since spoken to several fishermen about my inability to distinguish one fish from another. Then one day a more perceptive fisherman told me shame-facedly that he has the same problem with birds. "Try as I might," he confided, "I cannot see any difference between one swallow and another." I could certainly sympathize with that.

It wasn't long after that when meeting new people I noticed that there's a pattern to how I distinguish one person from another. In arriving at a new location, everyone looks more alike than different for a day or so. (They largely look alike because they are all strangers.) In short order, however, I can distinguish Jane from Anita, who no longer looks anything like Sheila to me. In no time at all, it puzzles me that other newcomers can't see the differences between them. Those differences seem so obvious to me—but only after association.

I will need to bump into Jane or Anita or Sheila once or twice before they even begin to become individuals to me. In the beginning, all I will notice is that they are all strangers. It will seem more significant that they are all around the same age, dressing in similar ways for similar employment, living in similar houses, in similar neighborhoods. And those similarities will seem far more significant than their differences. But quite soon, I will notice that Sheila is cheery at all times, no matter what is going on in her personal life. And by the time I distinguish her from the two other women I see her with all the time, I will be noticing all their differences. And I will no longer notice that they are all similar ages, that they all dress largely alike, that they all drive the same sorts of cars, etc. I will associate them by my own familiarity with them. And after associating with them, their differences will seem far more significant than their similarities.

People look far more like one another than different from each other. A friend who dabbles in cartooning proved this to me. First, he sketched a very simple face, then traced a second picture from the first, moving the nose a bit when filling in details. The result looked like an entirely different cartoon. He proved this again and again by moving the eyes then the mouth. And the overall appearance changed each time. (See accompanying sketch.)

Since then, I've noticed that individuality between one person and the next is further buried by fashion or fad. And the ability to distinguish one person from another depends on my ability to directly compare one individual to another. Geographical differences of hair color, eye shape, and skull shape are often dramatic, but only when taken from the extremes. Between those extremes, there is a great deal of minimal difference. Even then, the extremes often need qualification by comparison: "Taller/shorter than *who*? Darker/lighter than *what*? Could that tone of skin be a tan or is it a Native American/Asian? Or a light African-American? Or an Australian Aborigine?" By far, most of us fall into the middle somewhere.

Like people, other animals have some geographical variation within species. The weasels native to northern North America turn to ermine in winter; the same weasels don't change color in winter if they were born in the south. (That remains true even if they are switched after birth.)

There are at least five varieties of gray squirrels on the North American continent, including albino strains and completely black strains. And while I can't prove it, I'll bet they can easily, visually, recognize any of the others as strangers, even if the individual in question is the same color as every other squirrel they personally know.

In that same way, the differences between one person and another seem grossly obvious. Over time, their differences become more important than their similarities. Many people have no real interest in distinguishing one animal from another similar animal. And such people are flabbergasted to learn that animals such as zebras appear as distinct individuals to still other people.

Familiarity is a large part of the distinction between one individual person and the next because we have a vested interest in those differences. Once I become familiar with another individual, I have no problem distinguishing her (or him) from any similar-looking person.

But that vested interest in telling one individual from the next does not apply to the rest of the animal kingdom for lots of people. When those people look at other animals, they focus on the similarities—not the differences. And what they perceive is that all other animals (of a given type) look alike. Superficially, this seems like no big deal. But buried under that assumption is the conviction that *only humans have individuality.*

Since this is a hidden belief for many people, it's hard to get at. But if you talk to any of the people doing field studies on groups of animals who all look alike to an outsider, you'll find that those observers who have a vested interest in distinguishing one lion from the next know a great deal about the individuals they are studying. Such people can even point out family resemblances, interconnections with a group, daily patterns of certain individuals—correctly, of course. Individuality exists in all animals from elephants to zebras, anteaters to gorillas, baboons to yaks. Of course they look more like each other than different to most of us. But again, those who study those animals can easily perceive their individuality, just the same as the rest of us distinguish between Jane and Anita and Sheila. Just the same.

This distinction of individualism also heavily applies to behavior. Most humans behave far more like each other than differently from one another. The similarities are vastly greater than the differences. And this is just as true in the rest of the animal kingdom.

I've often heard that "'all starlings' (or) 'all robins' (or) 'all swallows' build identical nests." Do they? Certainly, the nests of a given species are far more alike than different, but the type of construction may seem more significant to us than any variation in material selection. It's possible that there's a high rate of individualism in material selection, even within a species. Many of those differences may be obscure to the people observing them, but obvious to the birds involved.

On the human side, I gained some insight into one possible distinction between one bird nest and another when visiting someone in a suburb. The directions to anyone's home get quite specific: "My house is the only yellow one on the street." In another instance, it was "We have the only cedar on that block"; still another time, "My house is the first one past the house with the flag pole, same side." Like birds, people distinguish their own homes by color, by a tree, by a precise location.

People also have a vested interest between one type of construction and another within the human community. Nearly anyone shown a picture of a pagoda will automatically identify it as a *Chinese* pagoda. The same sort of geographical identification is automatically added to *Grecian* columns or *Roman* arches. And that sort of thing makes it obvious that we've got predictable trends in our own architecture.

It's impossible to think of architecture and animals at the same time without mentioning that most famous architect: the beaver of North America. For all our tendencies to assume identical behavior patterns in other animals, it's fairly well-known that beavers build individual dams; no two are exactly alike. Beaver dams are designed *for* beavers and *by* beavers, but always to suit the exact location. Some dams bow upstream, some dams bow downstream, still others are built straight across a stream. And even the building materials vary from dam to dam, from beaver to beaver. I cannot say, "I know this is a beaver dam, because there are ninety-seven two inch logs in it; twenty-four are willow, fifty-three are maple . . ." etc.

One authority, impressed by beaver dams but puzzled by their individuality, finally dismissed the issue by pointing out that beavers will continue such activities even when caged and there is no logical reason for them to do such work. But I've met a retired builder or two. There isn't any logical reason for them to be building any more, either. Yet the one keeps fussing with additions to his house; the other has begun building furniture. Both of them (and many others) are trying to find logical outlets for their craving to build things.

If there actually is any fundamental difference between a beaver's engineering techniques and human architecture, it would seem to be instinct. Beavers obviously have a stronger drive to build things than does the average person. Yet once a person's innate drive to build is touched off by inclination and eventually compounded by education, the end result is the same. The beavers keep their young at home for a few years, during which those young help with ongoing house repairs—which strikes me as analogous to our educational systems. In the sense that basic beaver behavior to build arises from instinct, then perhaps *Castor canadensis* (beavers) have an easier time than *Homo sapiens* (people) do in fulfilling a natural desire.

People certainly build structures far more complex than the average beaver dam. But when I try to look at this with less human bias, it occurs to me that the complexities of our structures may be a result (rather than a cause) of needing to compensate for our lack of instinct with education.

Instinctual behavior in other species isn't as precise as the average person believes. While beavers build dams, the dams remain individual, from one beaver to the next; nest-building birds vary selection of nesting materials. Much of animal behavior is highly individual. This reality is easily missed or forgotten by people; we tend to think of animal behavior as inflexible. And sooner or later, those of us who regularly observe other animals get snagged on the realization that we've been viewing the world through the human prism of stereotyping.

When the great essayist E. B. White wrote an eloquent treatise in the mid-1950s about how a raccoon descends a tree, he believed he was describing how *all* raccoons descend *all* trees. It was six years before he witnessed another raccoon descend the same tree in a completely different manner. He amended his first essay with a footnote about the second raccoon's method. Like the rest of us, Mr. White had been caught up in viewing raccoon behavior through the human prism.

Some years later, author John D. MacDonald casually described the incision cats make in a mouse's abdomen (to remove the gall bladder) before the cat will eat the mouse. I had eight cats sharing my life at the time I read his description. I had never seen anything such as he described, nor have I seen any cat do such a thing since then. But both his cats did it all the time.

A good friend always shares her household with a few cats. She regularly strokes their stomachs. Without hesitation, she approaches cats unknown to her and strokes them, too. I watched her do this for years without getting scratched. I currently live with two cats and cannot (safely) stroke their stomachs—but my friend can. It is not simply being gentle, I assure you. I honestly think she can do it because she expects cats to let her do it. I don't expect to get away with it, and I don't; I get scratched nearly every time.

Whether the subject was appearance or behavior, I finally realized that when I looked for sameness, it was there. It was easy to homogenize other animals in both appearance and behavior. But once I began to observe their individuality, I could then see their differences more easily than I could see their similarities. In addition to that, I've heard more than one professional lament that she wished other animals were at least half as predictable as people believe them to be.

Beliefs (as opposed to facts) about other sentient beings are common all around the world. Still, it was startling to realize there's a pattern to the kind of beliefs themselves. As people in more settled areas of the world become further distanced from the animal kingdom by urbanization, our beliefs evolve into greater generalities. Not too long ago, I'd have been expecting a death in my family if a hoot owl landed near my house. During the same time, I'd have kept my hair short for fear a bat might become entangled in it. But even then, I'd like to think I'd have laughed if anyone had insisted that milk snakes steal milk from cows during the night.

While those particular beliefs have become passé, new beliefs rushed into the gap left by proving those early beliefs false. While modern beliefs sound a little less primitive, a little more educated, and a good deal more intellectual, they're essentially the same in many ways. And they rest firmly on the human prism of stereotyping, as the old ones did.

I recall an erudite gentleman who carefully explained to me that the *real* difference between people and animals was that animals lacked self-awareness. As this guy had put me in my proper place before, I asked him for his definition of self-awareness. His description sounded to me like self-consciousness—sort of how I'd feel if I arrived at a dress ball in blue jeans. When I pointed this out to him, he quickly dropped his air of delicacy and said bluntly "What I meant was that animals will defecate and copulate in the presence of others." On that note, he swaggered away.

This new belief—lack of self-awareness in other animals—seems yet another example of our peculiar avoidance of direct comparisons between all members of the animal world. It just so happens it's easy to come up with examples of people who will defecate and/or copulate in the presence of others. There are many people who have experienced being ill and have had to defecate in the presence of medical personnel. For myself, I freely leave the bathroom door open when I am the only human at home. I don't even think about my cat or dog's presence. And I frankly doubt I'm an isolated example. Pornographic movies never lack for stars. While the actors are copulating in the presence of directors, producers, and camera-operators, they are also fully aware that other people, whom they do not know, will watch the resulting film. And finally, *voyeurism* is fairly common, after all.

If human superiority is based only on the idea that we are self-aware and other animals are not, we've been seriously deluding ourselves for centuries. Simply put, self-awareness is just that: that "I" am an individual, distinctly separate from other individuals of all kinds. (Are we silly enough to believe only humans know that?)

Other animals who copulate and defecate in the presence of others simply don't anticipate any interference, anymore than I anticipate interference when I'm in the bathroom with my cat or my dog around. And it's probably the same subconscious conviction on the part of the X-rated movie stars, or the ill person in the hospital, or the domestic animal in confinement. No interference is expected, so why worry about it?

That's only a small part of the discrepancy of this assumed difference. Domestic animals will copulate and/or defecate in the presence of people but wild animals, under conditions of freedom, generally do not—and definitely not if they are aware of the presence of an alien observer.

Importantly, the very belief itself (that such inhibitions should exist) may be one hundred percent human in origin. Other species probably have inhibitions, possibly of sorts that we don't even suspect.

The modern belief of self-awareness as exclusively human is not the only new belief we've arrived at without much evidence. One of the latest beliefs has even become popular in some scientific circles: "For animals, there is only 'now.'" Skipping the awkwardness inherent in this statement, that science classifies people as animals (but they don't mean it here), there seems some secret meaning to this, like Bobo's fully opposed thumb. If the intended meaning is the obvious one, that *animals* don't have any sense of time, that *animals* don't remember things from one minute or season to the next, that *animals* aren't aware of tomorrow, it's still not a credible idea. As we can't truly communicate across the boundary of speciation, proof (either way) won't happen until another day. Meantime, however, there's at least a considerable body of contradicting evidence.

The famous story of Lassie, the collie dog who waited at the same time for her young master's school bus to arrive each day is not a myth. Anyone who shares her (or his) life with pets already knows they eagerly await your arrival each day. And farm animals anticipate being fed, watered, and milked at certain times each day, as any stock manager knows.

When I teach my dog to sit as a puppy, she obviously knows what I want when I next ask her to sit. Certainly, I continue to reinforce her understanding of the word throughout her life, but I can only reinforce something that is already there. (I do not have to re-teach the command each time.) When my cat becomes acquainted with someone new, she greets them as familiar (a completely differ-ent ceremony), even if she doesn't see them again for awhile. Many birds use the same nesting site year after year; might that mean they remember those sites from other years? When predators (like lions, tigers, and wolves) make a kill, they'll often leave the kill after feed-ing and return to it regularly for days on end. To me, that's substan-tial evidence there's a bit more to their existence than only *now*.

At least millions of mammals and probably billions of birds use the same migration routes year after year. It only seems reasonable to assume that some members of a given herd or flock recall the route from past years, that older members of a given group look for-ward to arriving at the same seasonal location they had enjoyed in previous years.

It also seems reasonable to assume from these few examples that other animals have some sense of time, memories of earlier times, and a sense of anticipation of times to come. Until people can truly communicate with other animals, many of our beliefs about how other animals behave are no more than whistling in the dark.

People have long had the habit of whistling in the dark about how other animals behave. The study of animal behavior appears to be centuries old. Books on the subject with ancient printing dates sug-gest as much. So I didn't expect to find that formal studies of ani-mal behavior are new. That makes it more understandable that many of our beliefs are a conglomerate of old and new information, form-ing together in sometimes odd combinations.

Until quite recently, field studies of how other animals behave in their natural environment were few in number. The oldest studies were often conducted by amateurs, many were superb at recording their observations. Some of these early studies accumulated data that implied that some animals have made substantial changes in their behavior patterns over time.

In spite of the acknowledged quality of many of these older studies, they are not usually the source of our accumulated impressions of animal behavior. The bulk of our modern impressions came to us, oddly, from zoos and laboratories.

Anyone with even a mild interest in animals knows perfectly well that an animal in confinement—pen, pasture, or prison—will behave quite differently from their kin who are not confined. The behavioral differences between confined and free animals are often dramatic.

The earliest behaviorists believed that any and all animals would behave in a certain way, no matter what the surroundings. ("Only humans have individuality.") This is where our modern concept of identical behavior in all animals of a given species originated—from these early studies done in zoos and laboratories.

Those early examiners were aiming at finding *patterns* in behavior. When I look for patterns, I see patterns. It's hardly surprising that the early behaviorists found patterns everywhere they looked: hyenas behaved in predictable patterns in this zoo and that zoo; so did lions; so did baboons. And so did every other animal whose behavior they studied. Voila! Look for patterns and patterns you'll find!

The reasons behind those early studies weren't very nice, either. They were godlike: their motive was to learn how to control animals to fill human needs. Superficially, this sounds negative, but there's merit in the approach. Consider, the man who tried to create a sense of aversion to sheep in coyotes. (Had this man succeeded, sheep ranchers would have no need to spread poisons around sheep-raising areas.)

Twenty years after Charles Darwin wrote *Origin of Species* and created such a public fuss, he suggested that there might be an evolution to an animal's behavior as well as to an animal's body. He was ignored by the scientists of his time; their conviction was that behavior should be studied for *patterns*. And it was already common knowledge that animal behavior was static, unchanging.

Although that common understanding didn't really disagree with Darwin's latest idea, Darwin's speculations implied that behavior might change in reaction to evolution. And that was the snag, because by then nearly everyone was convinced that animal behavior was wooden and predictable.

Because of the quality of the suggester, not everyone ignored Darwin's theory. A few people paid attention. As their numbers grew, this new way of looking at behavior became known as *ethology*.

As ethology became more popular, an intellectual war developed between behaviorists and ethologists. Behaviorists insisted (and proved) that they could accomplish amazing feats with animals subjected to their methods. Ethologists continued to insist that studies of behavior outside an animal's natural environment were less significant, lacking in substance, and had little merit.

As these two diverse groups argued, something unexpected changed. Ethologists pointed out that behaviorists lacked comprehension of why an animal responded in a certain way. In turn, the behaviorist wanted the ethologist to use controls to prove field observations. The ethologist wanted the behaviorist to acknowledge that an animal's behavior was vastly different in his natural habitat outside the laboratory and that the difference was important.

Each side eventually recognized its opposite had something to offer, but it wasn't until the mid-1960s that these two groups finally worked out their differences and formed the first interdisciplinary group, The Animal Behavior Society (ABS) from which everyone benefited. It's this new group that has formal associations with other sciences, from anthropology to zoology, and a great many others between these extremes.

In formalizing this new approach to animal behavior, everything would have been made easier if everyone could have begun from scratch, if all the earlier work from the extreme views could have been discarded. While much of that earlier work (from both sides) has little merit, it isn't simple to sort out which has merit and which does not. And much of that oldest work that seems to have the most merit is that of the earliest observers—the amateurs.

Importantly, some of that earliest work indicates substantial alterations in the behavior of some animals over time—behavioral evolution. One good example of such a behavioral change would be in the sea otter.

In 1741, an expedition headed by Vitus Jonassen Bering had the misfortune to hit a severe storm off the coast of Alaska. The ship crashed on the rocks, leaving the entire Bering expedition stranded on an island for quite awhile. Captain Bering died there, and the strait bears his name today.

Captain Bering had brought a naturalist along on that expedition, Georg Wilhelm Steller—*the* Steller who was one of the last people to actually see a living sea cow, the species that eventually became known as *Steller's Sea Cow*. These animals have been extinct for a long time now. The species was already becoming extinct when Steller first saw them.

To everyone's astonishment, Steller eventually returned from that expedition. His notes contained the only formal observations of living sea cows. The notes also contained details on the sea otter, an animal little known to people outside the immediate area. Steller also brought back more than seven hundred otter skins which, by direct implication, meant there were a substantial number of otters to be trapped. And that last caused the most excitement.

News of those sleek furs spread rapidly. Shortly, four nations became directly involved in the decimation of the sea otter population: Great Britain, Japan, Russia, and the United States. As early as the mid–1700s, a single trapper brought out more than ten thousand pelts. Sea otter fur quickly became an international fashion craze. By the mid–1880s, the traditional otter trappers—the Aleutian Island natives—were only able to trap 2,262 in a single year, a fraction of their usual number. Less than twenty-five years later, the situation was much worse: those natives were only able to trap 724. Clearly, the species was in trouble.

In 1911 the destruction of the sea otter population was formally acknowledged. All four nations agreed that the sea otter was on its way to extinction. Unless they all acted at once, there would soon be no sea otters at all. In an effort to stop the trend, all four nations signed an agreement to stop the decimation of the remaining otters. All four nations abided by that treaty, even when at war with each other.

The otter's numerical comeback was rapid and substantial. As early as 1938, a large group of otters returned to the California coast, where none had been seen for nearly a century. Their numbers continued to climb.

The sea otter represents a fine example of the successful efforts of millions of people. Although it is not the only example of such success, the point here has to do with the sea otters themselves. Those animals didn't just sit on their paws while enthusiastic trappers were exerting all the initial (and horrendous) pressure. They were reacting to those pressures with dramatic alterations of their basic behavior.

Major behavioral changes were noted as they occurred because of the earlier notes of Steller and a few others. As an example, Steller had written that sea otters commonly traveled about in family groups, that they seemed to be monogamous, that they rested frequently on shore in their travels, and that they had a *distinct breeding season.*

Today, sea otters are no longer commonly seen in family groups; they are now polygamous; they come to shore rarely; they give birth at any time of the year.

Since Steller's notes, no one has observed any specific breeding season among sea otters. It has been said repeatedly that Steller must have witnessed several breedings and assumed that those activities were examples of an actual (though nonexistent) breeding season. Bottom line: this particular observation by Steller has consistently been dismissed as incorrect. However, there is tacit acceptance of all his other observations. It is also understood that the entire species was put under sudden, extreme pressure, to force the animals into rapid behavioral changes.

Personally, I believe Steller was right about the breeding season for otters. My reasons are simple: it makes no sense for a cold-climate animal such as the sea otter to give birth at any time of the year. The young from such random matings would be in danger of freezing to death; their dams would have to scramble to keep them fed in colder weather; the stress the weather placed on both the young and the dam would be unnecessary and unusual among animals. And lastly, most cold-climate animals breed only at certain times of the year to assure the arrival of the young in the best possible weather.

Had this been the only behavioral difference noted by Steller, it would seem he might have been wrong. But other behavioral changes were noted *as they occurred* by other observers. During the end of the time of greatest pressure, the sea otters had already stopped practicing monogamy: more than one observer expressed surprise to see otters traveling singly more often than in the usual pairs or families.

An early sea captain, a Captain Hooper, was the first to note that otters no longer went to shore to rest or sleep. The otters had learned to sleep in the sea. They kept themselves from drifting by looping kelp around their bodies or by holding kelp between their forepaws as they slept. In this way, there was little chance they would drift to shore or out to sea—both being areas where a careless otter might be attacked by predators. Female otters even learned to give birth in the sea—quite a trick for an air-breathing animal (though not unique to the otter: whales do it all the time). All of this was new behavior in an animal who, not too far back, came to land regularly, even for naps.

Whether I'm right about that one point or not, it would seem that, overall, Darwin was right: basic behavior apparently evolves. The sea otter was under terrible pressure during that time. The only way to survive was to make rapid modification in standard (or traditional) behavior.

The likely reasons for these dramatic behavioral changes spreading so quickly from one sea otter to the next probably occurred in the simplest way: family groups were easier for trappers to catch and kill; the occasional family member escaped. And the escapee had a choice: either become polygamous or extinct. The young born of such impromptu matings had no family unit to remind them of their monogamous traits, and polygamy made breeding possible in such a hostile environment, the environment created on land by enthusiastic trappers.

Births at a regular time of year would have made the females and young vulnerable to trappers, and random matings tended to reduce those pressures. (Similar pressures on the northern fur seal in the Pribilof Islands may eventually cause such behavioral changes in that seal.) For the sea otter to survive under such great pressure, their natural inclination to a distinct breeding season had to be abandoned. Yes, people can indeed take credit for saving the sea otter species, but I think the otters' behavioral changes probably helped us save them.

The sea otter may not be the only animal reacting to new pressures. The woodchuck[13] (*Marmota momax*) may be doing something similar, though it isn't yet clear. Largely an animal of the eastern half of the North American continent, woodchucks are roughly the size of a large domestic cat, weighing between ten and fifteen pounds. They are most frequently reddish-brown in color, though I have seen completely silver ones.

It is the woodchuck's behavior that brings him to the attention of people. Woodchucks dig extensive burrows in hayfields and pastures. While always hunted for their meat, 'chucks are more widely hunted today because new farming techniques have made their burrowing practices a nuisance. Farmers curse these animals loudly and regularly because the burrows and accompanying dirt mounds cause difficulties with heavy modern machinery. 'Chucks clear pathways through planted fields, munching their way along to keep their paths open for daily use. Since modern farming and woodchuck behavior don't mix, the pressures on the 'chuck population have increased in the last half-century.

Standard woodchuck behavior will help explain the changes under way now. Though not generally a social animal, there is often more than one 'chuck in any field. And if any of them hear an odd noise, they will sit up on their haunches to look around. If the 'chuck sees anything dangerous, he will give a sharp whistle, followed by a series of shorter whistles as he runs for his burrow. But his initial whistle gets all the nearby 'chucks sitting up and looking around. (In some locales, the woodchuck is called a "whistlepig" because of this habit.)

As the 'chuck sits up on his haunches, he makes a clearly visible target. Fairly tall in this position, his reddish-brown fur clashes with the green background of field grasses. And he will sit up in this position whenever he hears a whistle.

Hunters discovered this behavioral trait long ago. Between that and the problems 'chucks create with new farm machinery, the pressure on the woodchuck has been increasing. The most visible among them were exterminated long ago, of course. But there were always a few 'chucks who dug their burrows on the edges of fields; these animals were far less visible when sitting up. Instead of a bright-green background, they were backed by woods or brush and their color blended into that background, making them much less visible to hunters.

Over the years, I've spoken to hunters from various areas where the woodchuck is common. All told me (without prompting) that more 'chucks are burrowing on field edges these days, far more than they ever did before. Older hunters invariably mentioned this. They also reported that 'chuck burrows used to be much more common out in the fields.

While the sea otter is an animal long-noted for intelligence, the woodchuck is not. Apparently, behavioral changes from environmental pressures may extend into any species—or we have long misrated the intelligence of woodchucks, which is quite possible.

I'm not trying to imply that such a behavioral change (unprovable and unclear as it is now) was in any way an intelligent decision made by woodchucks. It simply becomes a survival reaction to environmental pressure, a reaction that works. If the 'chucks who continue to burrow out in the fields die more frequently, they reproduce less often. The more unusual (at one time) choice of digging the home burrow on a field's edge simply results in woodchucks who reproduce more often. To the following generations, field edges might simply look more like *home* when it comes time to dig their own burrows. In this way, behavioral evolution might be occurring in direct relation to environmental pressures.

While woodchucks don't appear in danger of extinction, this small difference in behavior may be a change in reaction to environmental pressures. (Many species cannot do this.) Logically carrying this a step or so further, if woodchuck hunters evolve new methods of sighting 'chucks along field edges, perhaps this species may eventually take up grazing at night.

And if woodchucks do make such a change from diurnal (day) activities to nocturnal (night) activities, bodily changes would follow: the 'chuck grazing at night would need improved nocturnal vision, and his eyes would adjust to that need. Perhaps such a change would be as simple as an increase in the number of rods already present in the eye. (Remember, the higher the number of rods, the greater the nocturnal vision.) Such a change, theoretical as it is, would be an example of evolutionary change in body structure in direct reaction to environmental pressures.

Darwin's report of 1879 expressed that animals' *behavior* was as subject to evolutionary change as were their *bodies*. His earlier theory of 1859 was that an animal's body evolves in reaction to environmental pressures. His second theory, that followed twenty years later, has never been tied to the first theory, as far as I was able to find.

Behavioral evolution has likely always been the herald of a following bodily change. Another possible example of such a change might be the differences between the wolf and the coyote. Although the wolf tends to be ecologically delicate (wolves cannot tolerate living in close proximity to people), the coyote tends to be more flexible behaviorally. The differences between the two in body structure are minimal: the wolf tends to have a heavier coat and a larger size than the coyote. But the major distinction between the two species is behavior.

In recent years, hundreds of thousands of hours have been spent in field studies of behavior where the animals are observed in their natural surroundings. Because of these studies, we are only just starting to understand the reasons behind some other animals' behavior. And some startlements have already been discovered.

When David Pilbeam studied baboon behavior, he discovered that wild baboons act nothing like caged baboons, and he carefully documented the differences. This was important work, because caged baboons had been used in foundation studies on primate aggression. David Pilbeam's study unraveled a mass of work that had been considered beyond question for many years.

Until Jane Goodall's study of chimpanzees in Africa began, it was unknown that chimpanzees sometimes catch and kill other animals for meat. Until Farley Mowat was sent into the Canadian wilds to learn why the wolves were killing so many more deer, it was not understood that wolves were not the problem; the new shortage of deer was caused by a variety of changes in hunting laws.[14]

The point here is that all of our past and current understanding of other animals' behavior is filled with questions. While these various studies go on, we quickly learn that much of what we believed yesterday about animal behavior was appalling nonsense. Today, what we think we know about many animals isn't yet so certain.

There's likely to be many, many changes in the near future, as a result of current studies. Some of these discoveries will affect the human family. As an example, it is slowly becoming clear that animals that normally live in groups do not live in the free-for-all environment we've always assumed. Such groupings are made up of individual societies, filled with social inhibitions and moral values, just as is our own society.

I am not simply implying that the rest of the animal kingdom is becoming moral as people are. I am instead stating, absolutely, that *our moral values likely originated in the animal kingdom.* Evidence is accumulating that we people are very like the rest of the kingdom in our morality.

To belabor the point: it isn't the animals who are changing. It's our study of them that's changing. As our societies vary in moral codes from one community to the next, from one geographic area to the next, from one nationality to the next, so do the moral codes between one species and the next. And as these codes are being discovered through field studies, they are becoming expected, a subtle change among the observers themselves. There is less surprise today when a field observer reports that "these animals don't tolerate such and such behavior." Just a short time ago, such behavioral restrictions would have gone unnoticed.

Whether we've noticed them before or are only noticing them now, moral restrictions are definitely present in many other species. And they are most abundant in those animals with the greatest capacity for harming their own kind.

Dominant wolves will not kill other wolves that have angered them, no matter how great the provocation. Corvine birds (like crows and jackdaws) never peck at each other's eyes—even in heated battle. Individual chimpanzees will not interfere with a copulating pair of chimpanzees, nor will they steal meat from each other—the most prized food.

It isn't yet possible to list detailed moral codes for many kinds of animals; these codes are, in most cases, just being uncovered. And I'm sure that many of them will not meet with human approval or fit the human definition of morality. But it is becoming obvious that morality, in its broadest sense, is by no means exclusive to people. It is also becoming obvious that behavioral restrictions are fewer in the supposedly gentler herbivores, and are most strongly represented in carnivores and omnivores.

We commonly see rage, greed, and other volatile passions in our house pets. We assume such emotions exist in the wild. But we also see, again in our pets, a surprising amount of polite behavior, bravery, and general decency. Too often, we dismiss such behavior as only a result of training. And to top off our skewed misconceptions, we dismiss any bad behavior as animal nature.

When the physiologist, Ivan Petrovich Pavlov, was able to prove the validity of conditioned reflexes in animals in the early 1900s, he used dogs for his experiments. But his work has been largely used in the field of human psychology, with good reason.

Essentially, his experiments proved that when a bell came to symbolize the arrival of food to a dog, the dog would begin to salivate in response to that bell's sound—in direct anticipation of the food expected to follow the sound. It has long been understood that there is no essential difference between a trained dog's response to that bell and a human child's response to a lunch bell.

Reflexes are common to all species. If the body form is similar: limbs can be made to jump on most animals who have them when a nearby joint is tapped; all living beings move away from a too-hot heat source; most entities react in one way or another to displays of aggression.

It will probably be eventually proven that many animals have an innate sense of morality, and that it's likely a form of conditioned reflex on their part, as it probably is on ours. There's little reason to assume we are any different.

Yes, our social rules are probably much more complex than that of any other animals. Their complexity probably increases in direct relation to our increasing population. But the essential parts of our morality are probably static, the same as when humans first evolved.

CHAPTER 8
THE BEAST WITHIN

*I*t isn't difficult for anyone to describe a caveman. By consensus, it is easy to visualize him as squat, hairy, lowbrowed. In the same scenario is his inevitable club, dragging from one long arm. The club is an important feature of this picture, as it's commonly believed to be an integral part of his eventual winning of supremacy in his violent world. Modern people, faintly embarrassed by his involvement in our ancestry, assume his tendency toward a vile temper with a short fuse. We assume he was a murderous little man—a beast.

All would agree that we have come a long way from that nasty little guy. Still, our innate instinct for violence crops up with monotonous regularity. Overcoming this beast within us seems to be the main goal of civilization. Caveman's dubious gift of a genetic tendency toward violence makes a good many awkward (at the least!) problems in modern times.

Our ancestors' inherent nastiness was uncovered long ago, and accepted as a part of our genetic makeup. Caveman's appalling tendency toward violence was first uncovered, appropriately, in a cave, just after the conclusion of World War II. No one thought the timing of the find in the least significant. Yet the timing of the find explains some points that nothing else does.

It was a *world* war. Few nations escaped direct involvement; those few nations who did escape being involved survived under the constant threat that at any moment the war itself might suddenly shift and that all-encompassing devastation, that menacing horror, might unexpectedly become an integral part of our daily existence. The many, many nations that were directly involved knew beyond any doubt that the nature of man was hideous beyond measure. The nature of man was suddenly beyond comprehension to most of the world. The nature of man was filled with astonishingly monstrous capacities never before suspected.

If it was only ten or a hundred or a thousand shocking deaths that convinced the individual of the hideous baseness of the nature of man, the individual's sense of disbelief was quickly bypassed, and the entire species of *Homo sapiens* was overwhelmed with this horror before it was over. Russia alone lost 20 million people, nearly a twelfth of the population; while the Jews lost more than 6 million people, the Gypsies, who had less people to start with, were very nearly exterminated. Overall, the world lost 55 million individuals.

During this ongoing disaster, belief in the basic quality of humankind was at an all-time low. As the war dragged on and the toll of horrific atrocities mounted, the disbelief of the individual was steadily overwhelmed. With all of this in mind, what happened next is more comprehensible.

Doctor Raymond Dart, an anatomist from Australia, made a discovery that tended to explain some of the terribleness of World War II. Dr. Dart was a well-respected professor at a major South African university in Johannesburg. He had achieved international fame in 1924 as the discoverer of the *Taung Baby*—the earliest hominid skull that had been found up to that time. His find had slowly been accepted as significant; his perceptive realization that the Taung Baby wasn't just another ape skull was eventually recognized as incredibly important.

Dr. Dart was unable to continue his work until *after* World War II. At that time, he found a cave in Makapansgat, 200 miles north of Johannesburg in South Africa. Over time, he sifted an enormous amount of debris and was able to isolate many fossilized bones. A small fraction of these bones were hominid. But also in his hard-won collection were forty-two smashed baboon skulls. Dr. Dart declared that the baboon skulls had been smashed by the weapons of early hunters. When he was unable to find weapons there to fill out this picture, Dr. Dart concluded that *the other bones* at the site were the actual weapons used. He derived an exciting explanation of this period of early man and referred to it as "osteo-donto-keratic"—directly translated as "bone-tooth-horn."

And there it was. Finally. It was scientific evidence, apparently, that the nature of man was definitely horrid beyond belief—and that the detestable nature of man was hidden in his animal origins. World War II had proven that this savage nature resurfaced again and again, even in the most civilized people. In the perspective of the time, the unexplainable horror and devastation suddenly had an explanation, something we could point at that made the rest of what we'd recently witnessed clear.

Less noticeable, less interesting but equally substantial was that the "osteodontokeratic" theory provided an absolute, unquestionable separation for people from the rest of the animal kingdom. There we were, definitely, firmly and finally separated from all other sentient beings by our base nature.

As an explanation, it had all the advantages: it explained the horrors of war, it was scientific, and it *separated* us. That helped to explain our rapid advance as a species: by natural inclination, we would murder any being—human or other—that stepped into our path. It made sense—perverse sense, but at least it was logical.

What happened next is best described by Donald Johanson and Maitland Edey, in *Lucy: The Beginnings of Humankind*:

It is doubly embarrassing for anthropology that Dart, while he was at the height of his osteodontokeratic concoctions, should have met a traveling American journalist, the late Robert Ardrey. Ardrey had a vivid imagination and was immediately taken by Dart's vision of ancestral beings whose frightful propensity for killing their own kind separated them from all other living creatures.

Mr. Ardrey was an exceptional writer. He admired Dr. Dart's ideas and made certain they received exposure. Not only did this exposure of Dart's ideas become highly popular, they also gave subtle renewed impetus to the old established war of *Man Against Nature*. Besides fighting nature in general, people now recognized and tried to overcome their own, internalized nature. The "osteodontokeratic" explanation of the nature of the beast within was definitely an idea whose time had come.

I mustn't forget about those crushed baboon skulls. After careful examination of thousands of piles of bones, it was eventually declared that Dr. Dart's pile was rather typical. The reality of those skulls never received nearly as much publicity.

This reality was ever-so-dull and didn't have anywhere near the glamour with which to launch a thousand books. The real explanation left people without an ancient hero. It implied that the hominid bones found by Dr. Dart were probably the leftovers of some leopard's meal instead of the remains of some mighty warrior who went around bashing in the skulls of near-relatives (baboons) for food. The truth suggested that early people had pathetic beginnings, constantly fleeing from stronger predators and (often) not escaping. Reality sorely lacked charisma.

Dr. Dart's view of the earliest people had focused on Australopithecus—the first hominid known. The earliest skull he had found (the Taung Baby) was that of a six-year-old child, the age being discernible by the eruption of the molars in the jaw. Even today, science doesn't really know a great deal about Australopithecus, who lived in pre-Stone-Age times. But since Dart's first find, a great deal has been learned about the following peoples.

It wasn't until the mid-1950s that any clear proof of difference between Dart's view of ancient people and reality was uncovered—literally. By that time, unfortunately, the idea of early man as a murderous monster was established around the civilized world. Though the proof that early people were quite different was well publicized, few outside of anthropological circles paid attention.

The proof was stronger than anyone might have hoped for. It was discovered that even before Cro-Magnon, as far back as Neanderthals, early people buried their dead.

Evolution is a slow process. Very little of it occurs in a brief span of time. Had the nature of man been anything like modern assumptions, those early deaths would have been surrounded with (at the least!) evidence of greater practicality: a body dumped over a cliff, perhaps signs of cannibalism, or at least abandonment of the sick or injured. After all, this is caveman being discussed here—the nasty little guy with the club. The delicate sensitivity of the burial of a loved one just doesn't fit his image at all.

We still bury our dead, so that gentility has carried through the many thousands of years from those ancient peoples. A burial today doesn't entail the incredible hardship it must have represented back then. It's only when I step (psychologically) into the bare footprints of those ancestors that I can even begin to understand that any burial in those times was the ultimate hardship. Without even the benefit of a steel shovel, the backbreaking chore of digging a hole must have taken an enormous amount of time and effort. Yet such graves have become a common find throughout Africa, Europe, and Asia. No one in anthropology is even surprised by them any more.

But the graves, while a surprise at first, weren't the biggest surprise. Those much-maligned ancients buried their dead on beds of flowers.

Their environment was hostile to a degree we can barely imagine. Their environment didn't have flowerbeds. The hunting and gathering of flowers in such a time meant the appalling possibility of an unexpected double burial. Still they took the time and risks involved to do just that.

Homo sapiens may not be the only animal to bury his dead. There's evidence that elephants share similar ideas. The involvement of flowers may be unique to people, however; a small, but nice difference.

By this time, we know a lot more about caveman. We know that he functioned as a part of a group. And, as was mentioned in the last chapter, all groups seem to govern their behavior by general rules. Our ancestors weren't exceptional in that way.

But they were exceptional among the animals in other ways. We've never discovered anything remotely like artistic tendencies in other animals—at least, not yet. And early people were high quality painters. Paintings found on cave walls indicate a delicate perception of detail. In fact, the earliest cave art is considered beautiful, even by today's standards. Such art was certainly not made by some insensitive clod.

When I look at caveman in this more sympathetic light, I see him quite differently. He is more comprehensible to me. I know that he cooked his food the same as I do today. In this more compassionate examination, I can perceive the wonders and excitements of his world: how exhilarating it must have been to learn to control fire, bring it safely into his home for cooking and keeping warm. Our ancient ancestors must have been bright, compassionate, creative people, else they could not have accomplished all the things they did accomplish. From their humble beginnings, people have become the main force in the world of today.

Once early people had begun on the track of creativity, they continued to succeed at an ever-expanding rate. Slowly, they discovered that various plants produced seeds, and that those seeds could be gathered, taken to a new location, put into the ground, and grown there. It took incredible ingenuity and careful observation of detail to realize that if nearby competitive plants were removed from the area of newly planted crops, the carefully selected plants would grow better. It's on the basis of these fundamental discoveries[15] that modern people have created agriculture capable of feeding the entire human race.

The creativity of people, starting as early as we know it did, continues. From this more sympathetic view of early people, I can see how difficult each achievement must have seemed at the beginning. Like the simple idea of an in-home water supply: the pragmatists of the time probably pointed out (quite rightly) that anyone who wanted a drink of water could just stroll to the stream (or pond or river) when a drink was desirable. Those pragmatists no doubt jeered at their contemporaries who went to extraordinary difficulty to find the right kind of clay, wash it, shape it, bake it in a fire, fill it with water, then lug it all the way back so they might have an independent water supply right at home.

It's as likely as not that the jeered-at individual was a cave *woman*. By long tradition, modern people tend to think of cave *man*, but realistically, cavewoman was the likely creator of a good many concepts that we're still building upon. She was probably trapped into taking care of one infant while heavily pregnant with another. And that reality makes it more likely it was she who decided to find some way to keep water at the cave, so she wouldn't have to lug that big-headed baby and herself back and forth to the water all day, every day.

She had to be clever, just to stay alive. Her offspring might have died with monotonous regularity had she not tried to heal them. The idea of saving animal hides for later use, rather than eating them as many animals do, probably stemmed from the need to cover that hairless baby with something to keep him warm. And all the creative extensions that came from that—clothes, blankets, tepees—are all based on that originally simple idea. The inventive creativity of people was long established before anything like written history came along. And that creative inclination, by every measure I can perceive, is strictly human.

In every possible way, all the more complex ideas of people are based on those simple early concepts of the ancient past: the wheel, control of fire, and the obscure idea of growing food. What unsung genius is back there somewhere who originated the idea that her (or his) family might be able to live in *any* location just by collecting stones and creating a new cave? Only in recent history can we recognize that some of our most significant achievements come to us from ancient times. The unlikely concept of piling stones on either side of an opening, then joining them at the top with a stone shaped just so—the arch—can be proven to have long been in existence as far back as 144 B.C., The astonishingly innovative idea of making stones where they were wanted—concrete—can date back even farther to 193 B.C.

We have continued to build, on top of all these basic ancestral concepts. Yet without a doubt, nearly all of our modern creativity has been added *on top* of the creative ideas of our ancestors. The most complex technique of modern medicine arose from the (originally) novel idea of trying to care for the sick. The most glamorous structures in the modern world are, in essence, based upon the idea of building an independent structure—and that structure is most likely resting on the ancient idea of concrete. Agriculture, art, domestication—basic ideas that can be traced back to the very earliest people.

Our creative inclination has tended to speed up over time. Success tends to do that. By now, we've accelerated beyond any other entity on Earth. We've piled achievement onto achievement. Early people might have taken generations to proceed from the notion of an independent water supply to the achievement of the actual goal. But modern achievements tend to accelerate each other at astonishing speeds.

Take the idea of flying, for example. Certainly, the concept itself is very old. But from the flight of Orville Wright of a 120 feet at Kitty Hawk, North Carolina, there is less than sixty-six years until Neil Armstrong took "one small step for man" on the surface of the moon.

Success tends to make incredible things possible.

From the lofty position of modern times, it's easy to assume that all the creations of early humanity took half of forever to come about. In the same way, there's little question in anyone's mind that all sorts of changes take place more rapidly in today's world. However, not all comparisons with ancient people make modern people look so good. Early people lived in a world fraught with immediate dangers: the constant possibility of being killed and eaten, or dying from a simple illness like a cold. The specter of starvation lurked around every corner as well.

Any superficial look at the modern world makes it seem like a safer place. Yet when I stand back a bit, I can see that all of the dangers our ancestors were concerned about were *immediate* and wholly dangerous to the *individual*. Our modern dangers are greater, frequently harming vast numbers of people; some modern dangers even threaten our *entire species*.

In the beginning of the twentieth century alone, more than 70 million people were exterminated in wars. The number lost in World War II alone—55 million—is probably vastly greater than the combined number of Neanderthals or Cro-Magnons or even Australopithecines who ever lived.

Of course, in modern times, there is a vast amount of crime with highly personal intent: husbands and wives murdering each other, parents killing their children, children killing their parents, siblings killing siblings. In addition to that, we have endless crimes committed by one person against another with depressing regularity: theft, rape, beatings, murders. And we also have a great number of crimes that seem more focused on the species itself, rather than on any qualities or nastiness of the victims: terrorism, slavery, bombings, warfare. Such crimes seem more focused on *Homo sapiens*. There is ample evidence everywhere that modern humanity hates modern humanity with an active passion. Just look around. Just read today's newspapers or watch the news on television.

For at least centuries, while we've been killing each other off in staggering numbers, we've also been holding (sometimes unannounced) wars against other species: the eagles, the hawks, the wolves, the bears, the lions, with a long etcetera after these few mentioned. No other species has been too big (like the whales) or too small (like the mosquitoes) to avoid our attention. Yet no other being has ever received the degree of hostility we've shown each other. Nearly any schoolchild can rattle off a list of wars where at least thousands—more likely millions—of people have been killed by other people.

The nature of man as we define it in modern terms would probably confuse the hell out of Cro-Magnon and Neanderthal people. It's quite likely that even Australopithecus, if he could have understood, would have been astonished and bewildered at Dr. Dart and Robert Ardrey's ideas of his behavior. Our history of hatred toward each other is long, but not that long. *The beast within*, as we now define him, is very likely a new invention, very now, and complex creation of thoroughly modern people.

Like every other creative concept, we've been building upon this one. There is little hesitation before some modern individuals build new additions of horror on the accumulating horrors of the recent past, a kind of "If you think that was horrid, wait'll you see this!"

Yet the evidence that our ancestors, with their inherent tendency toward creativity, compassion, and possibly kindness still shows itself through us with reliable regularity. The evidence that we do care about each other—and always have—continues to outweigh the mass accumulation of evidence of our monstrous modern tendencies. For every ounce of proof that many modern people do not care about anyone beyond the tip of their nose, there is a hefty pound of evidence to the contrary.

In general, evolution is not an overnight thing. People still scream for assistance when attacked. If assistance were not a real possibility, why bother? Unless, of course, it is really possible to alert help. If there were no real chance of alerting another to our personal peril, we could simply faint as the opossum does, or flee as the robin does, or try to rapidly dig underground as the mole does. The scream is an automatic reaction in search of assistance. It has long been with us, and will likely be with us into the far future because it works so often.

No matter how hard times have gotten, the sacrifices one person is willing to make for another person always bob to the surface over the mire of immediate trouble. Always. By far, the worst and best examples of our extreme behavior in humankind surfaced again and again in World War II: when food was so incredibly scarce in German concentration camps that possession of a small morsel might mean life itself for yet another day. There were many examples of food given away to those who needed it more. When a place in line meant possible survival of a gas chamber, individuals changed places with those farther ahead, directly trading life for death. People risked their lives each and every day to help still others escape. Many of them died doing this sort of thing, proving that dangers were real.

In spite of the modern conviction that humanity has been changing toward the worst since those awful times of a half-century ago, evidence is abundant that no change has taken place in our essentially decent, caring nature. Every day, in every nation, people run back into burning buildings to save others trapped within them; still others dive into unknown waters to save strangers who appear to be drowning. And there are always abundant volunteers for search and rescue groups when people get lost in snowstorms or when children are missing. These significant compassions of ours are as reliable as the arrival of another morning.

Has our basic nature changed in the fifty-three years since that war ended? Behavior is about the only thing that could make such a speedy, dramatic change in response to evolutionary pressure, as I pointed out earlier. But there isn't much evidence of any basic change. If we were less sympathetic toward each other, people would never applaud the success of any other person. Realistically, there is no benefit to the people clapping in any situation like this. But we do applaud the success of others.

We applaud thunderously.

The better side of the nature of man is everywhere apparent. Another example would be the polygraph test, commonly called the lie detector test. It speaks volumes for the fundamental decency of people.

Ex-policeman Douglas Gene Williams explains how to pass a polygraph test on the basis of how the instrument works:

> I can tell the complete truth, or a complete lie, or anything in between and still pass any lie detector test given by anyone, any-where. The polygraph test is not a lie detector, and it is not a truth verifier. The polygraph is simply a crude reaction recorder, and the reactions it records can be indicative of just about anything.

Mr. Williams goes on to explain that complete calm, no matter what question is asked, is necessary. Calm breathing gives deceptive readings to such a machine. The machine is simply used to record tension. Incorrect readings can be caused by breathing faster, tightening of the anal sphincter, even tensing any muscle unexpectedly.

Whether or not anyone can pass such a test while lying is essentially irrelevant to the main issue here. The premise behind the polygraph rests on the concept that people are so honest that those who don't tell the truth will have *a physical reaction to telling a lie*. What more simple statement can there be that people have an inherent habit of honesty?

There is proof of our natural compassion toward each other, our fundamental decency. In any city in the world, people will pass casually under steelworkers strolling along metal beams high overhead. If passersby notice those workers at all, it's usually just a passing observation: "You'd never catch me up there!" Generally, no one pays much attention to such things on a conscious level; most of the people passing through such a scene will barely notice.

But the entire situation for everyone there changes instantly to quivering alertness if the steelworker should shout or slip. All of those who barely noticed his location will look directly at him. Every observer will feel a tingling sensation as the hairs rise on the backs of necks. And if the steelworker should actually fall, many of those passersby will vomit spontaneously; others will simply feel nauseous all day.

The reliability of these reactions speak of centuries, millennia, eons of mutual care, concern, and involvement in each other's fate. These reliable reactions will occur in the smallest villages or the largest cities.

Examples of the essential nature of man are easy to come by: every year, all around the world, thousands of people donate blood to hospitals for little (if any) personal gain. By far, most donors have no idea who will receive their gift of blood. It's generally handled anonymously.

Some years ago, donors in Great Britain were surveyed as to why they chose to donate blood. Less than 2 percent of those donors gave extraneous reasons for their donations, such as "it makes me feel better physically to get rid of excess blood." Most insisted they did it for the general good of people.

Only the more complex animals have anything remotely like our consistent mutual concern. Our inclination to sympathize with each other holds true, whether the people involved are friends, relations, or complete strangers. Our *kind* of concern isn't unique, but the *degree* of our concern is certainly unique. The majority of other animals seem to have lesser responses to others' good or bad life events.

We haven't yet reached the point where we can prove that our degree of concern for each other is greater than mutual concern is among other animals. However, there is some evidence in our social behavior that says a great deal about the high caliber of our mutual concern.

As of now, we know that the breaking of social codes among the animals results in punishment every time: among wolves, crossing the carefully demarcated boundaries of another pack, a non-pack wolf is chased out *immediately*; the hen who steps out of the pecking order to snatch a piece of grain is attacked *immediately*; the young stallion who tries to breed the mare of another stallion is attacked *immediately*. Punishment is swift among most social animals.

It's only slightly different in the human world. We seem to have more social rules than any other species, probably because of the complexity of our social groupings—and the density of our population. Because of this reality, punishment is often a delayed activity. This must have been the case far into our history, because we have developed this odd tendency to feel guilt about any infraction of social rules. Guilt seems to be rare among the rest of our kingdom, quite probably because punishment is usually so immediate in their societies. True, the dog is another animal that clearly suffers the emotion of guilt. Dogs have been our close associates for so long, it seems certain we've given them this dubious gift.

Like the dog, we genuinely suffer from guilt, which we most often refer to as our conscience. Even those people who claim they don't have a conscience will usually admit to an occasional twinge of guilt. (Erma Bombeck, the writer, once said, "Guilt is the gift that keeps on giving!")

Why should anyone feel guilty if someone else—a stranger—doesn't approve of his actions? If people really didn't care about each other, if people didn't seek each other's approval or admiration, guilt would serve no function. But we do feel profound guilt when we have done something that we know other people would not approve. Where does this feeling originate?

The beast within is an ancient animal, overwhelmed with compassion and sympathy toward all fellow beings. And no matter how much pressure we put on this beast, he forces us to think about our actions, to notice the reactions of the other beasts, to look out for quite a lot more than number one. Our sympathies are aroused by a variety of events that often bear no possible personal overtones.

Always, without fail, we stand up for the victim. If people see a lion running down a deer, our sympathies are automatically aroused by the deer—not the lion. (If our nature were truly predacious, wouldn't we cheer the predator?) If we hear of a child who has been kidnapped, we automatically feel sympathy for the child. (I have never heard, "That poor kidnapper! The child will be such a nuisance!") When we hear of a blind person being robbed, we automatically feel anger toward the robber.

In any of these cases, why are our sympathies aroused? People everywhere know perfectly well that thousands of lions run down thousands of deer every day; people know perfectly well that an appalling number of kidnappings occur regularly; and a reliable percentage of robbers consider the blind natural victims. What possible difference could it make to know that still another example of any of these events has occurred, especially since everyone knows perfectly well that such events are ongoing, normal, downright mundane activities?

It is the beast within who cares. It is the beast within who is still fundamentally an animal from a small group, a village mentality, who has a compassionate streak that is the proverbial mile wide. It is that beast who knows that the fate of any life is heavily bound to the fate of life in general. It is that beast who howls through time to thoroughly modern people that what can happen to that deer, what can happen to that child, what can happen to that blind person can happen to any of us. And we care. Deeply.

It is the village mentality of the beast within who ignores the modern fact that the world is full of sadness and horrible incidents, which in no way diminishes his (or her) genuine sympathy toward events personally witnessed. It is the beast within who demands we feel compassion toward any event personally witnessed. It is the beast within who demands we feel something about every sadness, every horror, and every wrongness we witness.

In this way, the beast within, be he Neanderthal or Cro-Magnon or both, is clearly alive and thriving in modern times—in spite of being overwhelmed by modern stress. In the deepest sense, he is a beautiful beast, not greatly different in his focus than most other animals. And he perceives things on the same scale.

While his goodness is everywhere apparent, there is an elusive, negative cost attached to his nature. Like most animals, he insists on focusing on the personal level of any event that catches his attention. And his capacity for compassion is restricted by the scope of his focus. His sympathetic circuits are easily overloaded.

The beast within doesn't have the capacity to see the larger picture of the nature of man. But if he could, he would see that the nature of man hasn't changed a smidgen since the beast's primordial time. For each individual the beast bumps into (in these modern times) who seems capable of stealing from the blind, kidnapping children, or other mayhem or murder, the beast will also bump into hundreds of people who are equally incapable of such actions.

With his mentality focused on the flock, the herd, the tribe, the beast's focus remains small, and he cannot see that he is continuously focusing on the exception and not the rule among his own kind. Like most beasts, he keeps his attention on those of his natural group, with only an occasional peek at another flock, herd or tribe.

Zoology has a good word for this way of seeing things: *extant*. And it applies here to the focus of the beast.

Ram sheep do not kill other ram sheep in dominance battles. Of course, there are extant individuals who do so. Rabbits do not kill other rabbits, except for a few. People do not kidnap children or rob from the blind or murder each other—with the exception of extant individuals in each case. And the number of extant individuals among people continues to climb, as does the number of people on the planet. (In fact, the percentage itself may rise, as individuals continue to react to the pressures from overpopulation.)

The nature of man and the beast within are one and the same animal. On a more collective (group) basis, both have a greater capacity for different sorts of behavior. But on an individual basis, the nature of man is no less noble than the nature of any of the other beasts.

Any overview of the rest of the animal kingdom makes it clear that people are not the only animals with such innate tendencies toward mutual concern, willingness to make genuine sacrifices for the good of the group, or offer genuine help to others in distress. A wide variety of other beings practice that kind of behavior all the time.

But in that same overview of animals, people look like pretty good animals, with (probably) a magnificent future. There's a lot to indicate that.

Look around!

CHAPTER 9
ANIMAL RIGHTS—AND WRONGS

*T*here's little question that people are currently the dominant animal on earth. Any thoughts on this subject usually begin with that reality firmly in place. But few seem to question why. The reasons we hold this position of dominance are more varied than may seem obvious.

Our dominance doesn't arise from our physical strength. Neanderthal people were far stronger than modern people. It might be nice to assume that as the brain developed, we needed strength less and slowly lost it as an unnecessary feature. There may even be some truth to that, but there's much more than simple brain development behind our domination.

Physical weakness can be costly. Throughout more recent evolution, people have become progressively weaker. The polar bear (*Ursus maritimus*) has size and strength over us. Such bears weigh many times what the average person weighs; they are also fast for their size—certainly speedy enough to overtake any fleeing human. But, unlike people, polar bears are limited to the frozen north; they are unable to travel very far south without technological assistance. People have always been able to live in a much wider variety of climates than the polar bear. In this precise way, we can conquer this much stronger and much faster animal.

Yet another way that people conquer polar bears is in sheer numbers: people vastly outnumber all the polar bears who ever lived. In these precise ways—climate and numbers—people have become dominant over polar bears. The bears' advantages of superior size, superior strength, and superior speed are only helpful to them under some circumstances.

Conversely, human advantages over rats (*Rattus rattus* or *Rattus norvegicus*) are quite different. So far, rats still outnumber people, though the numbers are probably closer than either species is comfortable with. Unlike polar bears, rats have fewer climatic restrictions; they travel readily with people to all possible climate extremes.

Unlike people, rats do not need to develop technological aids to travel to extreme climates; we people develop those aids and simply take the rats along. They can rely on us to work out the details, and we can rely on them to adjust to whatever limitations we impose on them. People may be physically stronger than rats, but we aren't nearly as adaptable. And in this, rats are more advanced than we are.

Human dominance is accomplished by simply slipping through the weak points of other animals' defenses. When it inevitably became clear that it was only sometimes possible for people to overpower polar bears with crude weapons, people were able to go south to lick their wounds while other, more powerful weapons were devised. Of course, some people have always stayed behind with the polar bears, maintaining an armed truce. And even under those more dangerous conditions, the people have a distinct advantage over the polar bears. That advantage isn't necessarily wit, nor weapons; while subtle, it is an advantage that is rare among animals: people will fight for each other; polar bears will not.

A mother lion (or nearly any other dam) will put up an earnest battle to save her cubs; a father wolf will do the same for his pups and other relations. But humans will put their own life on the line for any other human—most especially in a battle with another species.

A human who stumbles upon another person in pitched battle with a lion or a wolf or even a domestic dog instantly assists the human involved—not the other animal. No one stops to assess the situation ambiguously: "Is his skin tone the same as mine? Are his eyes shaped the same? Is this person someone I know and like? What are his political leanings?" Later, such questions may arise. But first and foremost, the human being embattled by another animal is human. Assistance will be an instantaneous response in nearly every case. (Without question, there have been a few extant examples where this hasn't occurred, but such examples have always been rare.)

The rat is such a successful life form because of adaptability, but he will not defend rats in general, anymore than the polar bear will defend another polar bear.

This has been a skewed truth at times. Not too long ago, it was common practice that if one lion killed one person, if one wolf killed one person, if one shark killed one person, all of humanity would focus on that type of animal (who did the initial killing) and try to wipe them out. Such practices were standard, automatic. Hence, many animals achieved quite a close brush with extinction before people realized this line of thought was dangerous to people.

The reality of this approach is woven throughout human history. Massive attempts to eliminate the tiger, the grizzly bear, and the wolf represent a framework to a long part of human history.

In light of that, it's no less than astonishing that other animals have developed inalienable rights, acknowledged by people on a global basis. Other animals have the right to expect reasonable treatment from people, perhaps not in every circumstance, but in a broad range of areas. As only one example, this event occurred in the United States, reported by Joe Skorupa in *Popular Mechanics:*

> Charles Gibbs was in Glacier National Park...when he spotted a grizzly mother with three cubs...His 40 photos indicate he got

within 50 yards of the grizzlies and followed them when they tried to move away from him...Most people...believe Gibbs probably bore some of the responsibility for the attack.

Mr. Gibbs paid the ultimate price for his lack of wisdom. But the mother grizzly, clearly identified by Mr. Gibbs' photos, was considered within her rights to defend her cubs from that fella with the camera. No one hunted her down, as people would have done not too far back in time.

In June of 1988, a four-year-old girl was playing on the shores of a lake in Englewood, Florida (U.S.), which an alligator took as an opportunity for an easy meal. The alligator was actively pursued, though those directly involved knew the child was already dead. The alligator was found shortly thereafter and killed.

These two examples are good reflections of each other; the law enforcement agencies involved acted consistently. Clearly, the adult bear was not pursued simply because she killed a man. It was accepted that the man bore "at least some responsibility" in his own death. The child did not.

If people are vulnerable as adults to raw power, children are our weakest points. Even if it could have been explained to that alligator why he was pursued and killed, I suspect he would have been astonished: he saw an opportunity for an easy meal and he took it. He felt in the power position. He was wrong. Death was the price paid for a serious error in judgment. In neither case did one sort of being have automatic rights over another sort of being.

A recent survey, reported in Robert Walters' column, found that more than half the people in the state of Montana were delighted—no exaggeration—that a wolf pack which had crossed over from Canada had since expanded to three packs of wolves. Nearly three-quarters of those surveyed expressed hope that the wolves would remain in the area.

This is a far cry from people who (not long ago) put price tags on the heads of wolves, lions, eagles—any species that was in the slightest way dangerous to people. We've come a long way in our attitudes over a short span of time. This granting of inalienable rights to whole species was a big step. It came about slowly. Today, all beings have some inalienable rights. There is no longer anything automatic in humans having rights over other animals, nor viceversa. Whether an incident involves a bear, an alligator, a man, or a child, the rights are granted to the first victim every time. Whatever being was most within his rights at the time is automatically granted the most consideration.

This brings up an even more basic issue: where do rights come from? Why does any being, human or other, have such a thing as rights? All human communities, developed or undeveloped, have rights. An injured man has an inalienable right to expect (at the very least) a show of concern from any passing human, whether that passing human is an interested party or not. It's universal among people, whether in the center of urbanization or far off in a swamp, desert, or mountain range, that anyone passing an injured human will assist if possible.

Stepping back a bit, *rights* seem to be an extension of concern. And, as was discussed earlier, concern itself seems to be an extension of intelligence. All across the animal kingdom, concern shows itself only among the more intelligent species. As a human being, my concern extends itself to every other human being. As a mere citizen of the world, I'm concerned about human enslavement, discrimination against any member of my species, injustice wherever I perceive it to be. I recognize that, while it might not be my personal neck in any particular noose now, it could just as well be my neck tomorrow. My neck is a human neck, as human as any neck in any noose of enslavement, discrimination, and injustice.

It's only quite recently that I've come to recognize that the neck of the jaguar, the neck of the monkey, or the neck of the wolf isn't all that different from my own neck.

Over the long battle of *Man Against Nature*, I began to realize that

"I" was winning and my losses would be terrible if I succeeded. As an appropriate reaction, I began to extend my concerns. It was now inappropriate to hunt and kill a bear just because she was defending her cubs. Ever the pragmatist, it was equally unacceptable to let an alligator kill your child, when I was aware that his realization that your child was easy prey made my own child vulnerable. I grant rights with equal justice. Rights in general arise from the human community. It's people in every case who extend rights or take them away.

There are, of course, a few exceptions. My dog guards my home against any and all beings—except me. She gives me the right to enter my own home. But she defends my home whether I am there or not, and grants no one but family members the right to enter. She readily surrenders this right when I am there. I can then pick just anyone I want, whether she trusts them or not, and let them into my home. But those same people don't have access when I'm not there.

Anyone who has tried to mount a horse who was unwilling to be mounted knows that the horse is denying the person a right. This applies whether or not the horse has been trained to accept riders, though the trained horse can be more easily convinced to surrender his right to refuse a rider.

Such rights exist throughout relations between people and domestic animals. The cow who isn't acquainted with the person trying to milk her resists being milked; at times, with force. A bull will not tolerate the presence of a human stranger among his cows, but is often tolerant of people he knows in the same circumstance. All the domestic animals make it clear that they will cooperate only with known people, will resist unknown people at every opportunity.

The same familiarity is often not necessary for people. People who like any dog generally say, with no qualification, that they like dogs—or horses or cats or cows, or the greatest generality of all: animals. Other beings tend to be far more selective. They like only specific other animals—and it is a right they insist upon exercising.

As example, when eighteen kittens were raised in isolation and each was given a rat as a playmate, none of the kittens would harm their personal companion, though three of the kittens did eventually become rat killers. Each kitten had developed a sense of family with one rat. Even as the kitten became an adult rat killer, he would never harm his personal companion. Once these cats had achieved the ability to distinguish one rat from another, each exercised his right to choose which rats were fair game and which rats were family.

Since the time of King Solomon, people have seen the rest of the animal kingdom as one big family—quite separate from people. This perception made it okay to war against other animals as a whole. It made it acceptable to commit cruelties against other animals in general. Yet there always remained a few animals (different kinds in different places) where this criterion didn't apply. For example, it's acceptable in China to eat the family dog; conversely, in the South Pacific, it's more acceptable to eat the family and spare the dog.[16]

Animal rights and wrongs have always been modified by time and place. In recent times, other animals have been granted more generalized rights. And the division of rights tends to reflect this new attitude: monkeys have more rights than rats, dogs have more rights than mice, and cats have more rights than rabbits.

But the popularity of the animal isn't the only consideration on the issue of rights. The majority of other beings have far more rights *outside* a laboratory than *inside* a laboratory, no matter which species they are.

Biomedical research is a broad term covering product safety testing, drug testing, and surgical procedures. (It's considered by many to be divisible between product safety and drug testing.) A growing number of people consider biomedical research to be the ultimate wrong against other beings. Ultimately, it's a subject guaranteed to make a goodly number of people squeamish, so brevity is important. But it's the area of the most activity today involving people and other animals' rights and it demonstrates some brand-new attitudes in humanity.

Right now, in the United States, this issue has achieved the level of out-and-out war between the two extremes, represented by a variety of groups. In trying to maintain an objective distance, I mailed a small gift check to various organizations on either side, requesting information. In each case, I specifically asked to be left off their mailing lists.

All the animal organizations promptly put me on their mailing lists. Some sent me further information, which began *Dear Member*. Most supplied me with continuous literature and rather ghastly photographs. The biomedical research people sent me the material I requested, then left me alone, as I'd requested.

Rather than awarding points for politeness or receptivity, I read all the materials sent to me. I received approximately four pounds of reading material. I followed most of the major concepts offered (by both sides) all the way to their inception to gain an overview: is this strictly an issue of practicality or does it legitimately involve morality?

It was a long way through, but eventually I was able to digest that overview. Somewhere in the depths of my reading, I became aware that the overall issue specifically involves the larger topic of this book: how much *like* other animals are we? How *different* are we?

The issue took on a new dimension when I heard a little girl interviewed on television. She said she loved her cat very much, but would readily surrender it to medical science to save her friend, who was ill. There is no arguing with this child. She simply stated her personal position. Child or not, she was entitled to her views, and also entitled to surrender her cat, should she choose to do so. Her cat has no say in the matter.

But what about her cat? If the situation were reversed would her cat surrender this little girl? It's a fair question.

The answer is no. Of course I cannot know what that little girl's personal cat thinks of her. And I'm totally incapable of communicating with that cat, so I shouldn't make such a flat statement.

But I can, with absolute confidence. My confidence is solid, because of that experiment I mentioned earlier about the eighteen kittens raised in isolation with rats. That experiment proved that those cats favored those particular rats on a personal level. Even those three kittens that grew into rat killers didn't hurt their companion rats. So, as odd as it may seem, I can answer with certainty: no, that little girl's cat would not turn her over to an outsider.

It occurred to me that this current issue is the first variation in long-established positions about other animals: the *no essential difference* folks are *against* biomedical research; the *no comparison possible* group is in *favor* of biomedical research. And that shift in position for both sides is new; it's the first notable move of either extreme for more than a hundred years.

The *no essential difference* folks are saying, in essence, that the issue is firmly in the area of human ethics, that the basic issue is whether or not it is right to perform biomedical research on relatives, succinctly stated by one writer as "a rat is a pig is a dog is a boy."

The *no comparison possible* group is saying, in essence, there is no ethical issue involved, that other animals are not enough like people to categorize the issue as ethical; hence, there isn't any ethical issue. But they modify their argument in a significant way: animals are enough like us *biologically* to justify their use in biomedical research. Fundamentally, it's *your* child or an anonymous Fido, to paraphrase another writer.

The median position has also shifted, along with the extremes, and is fairly close to matching that last position. The actions of biomedical researchers are understood as an ugly necessity, which everyone readily admits is *situational ethics*. (Not nice, not fair, but necessary.)

Ethics are highly personal. The entire subject is an emotional minefield. It wasn't long before I realized that my idea of what someone else's ethical position should be is irrelevant. It seems to come down to whether human behavior in this area is justified by results. I found it difficult to find my way through the heavy emotions and cool (sometimes downright cold) logic to some simple answers.

According to one animal rights group, in the United States alone 14 million animals are killed each year in product safety testing. While the biomedical people question that figure, they offer no figures with which to make comparisons.

So I checked. *It's the law* in the United States that products pass a series of safety tests on other animals before those products can be marketed to humans. In this case, the law itself may represent what one writer referred to as "bureaucratic inertia," as there are many products which could easily be tested more safely (and with better results) through other testing methods. Animal tests are currently required by the Food & Drug Administration (FDA). Importantly, those required tests ignore a good many tomes written which show their record of past failures. (And some of those failures were huge.)

Fair, however, is fair. I wouldn't be in the least cheerful about a surgeon who told me he was nervous because it was his first day on the job. I'd like all the assurance he could offer me that he'd performed the operations previously. Many times, I would hope.

The biomedical research folks provided me with an impressive list of modern medical successes they say are directly attributable to such research. I'm not a casual reader. I checked. The Thalidomide disaster (a drug, heavily tested on other animals before being rated safe for human consumption, caused 65,000 people to be born with serious and sometimes horrendous birth defects) is swept under the rug. Instead, the success of the polio vaccine is offered as a perfect success story. But was it?

More than a million rhesus monkeys were used in experiments

186

to develop a vaccine against polio. When India and several other nations banned the export of these monkeys, some of the standard arguments for biomedical research were apparently used to increase the supply: "Are you *really* going to choose a monkey's life over the lives of human children?" (Your child or Fido.) When India relented, the consumption of rhesus monkeys soared.

Did all those monkeys surrender their lives to a good cause? Without question. Were their deaths useful in the end? Perhaps some—but definitely not all. The vaccine was eventually produced from a cell culture—the same alternative now available for much biomedical research and most product safety testing.

The biggest question of this issue is identical to the main theme of this book, as was mentioned before: how *like* other animals are we? How *different* are we? And isolating the question to strict biology doesn't make the matter any less complex. Awhile back, I mentioned Dr. Jenner's discovery of the smallpox vaccine. Wouldn't that example serve as a success story of biomedical research succeeding mightily in human medicine?

I'm not qualified to get technical here, the best I can do is simplify the issues. Some diseases cross the line of speciation. Some diseases do not. But in some cases, the diseases themselves are closely related to one another. Dr. Jenner did not introduce smallpox to the cow; he introduced cowpox (a related disease) to the people. On an infectious level, cowpox is an attenuated version of the deadly human smallpox. Had the cow been injected with the human smallpox, the biological situation might have been very different.

Another example would be malaria, a disease transmitted to people by mosquitoes. It makes humans quite ill. But it has no effect on the mosquitoes. Conversely, penicillin is given to people to control bacterial infections; the same drug administered to a laboratory guinea pig will result in death. But rabies, a terrifying disease, is essentially the same in all mammals.

This reality—that some diseases cross the speciation boundary and some do not—is well understood throughout biomedical research. Consequently, a great many exhaustive tests are performed on other animals before they can be safely used to create human vaccines. All other beings potentially carry viruses that may (or may not) be contagious to people. And nearly all of the time, such thorough testing is successful.

But not always. In one well-documented incident many years ago, a monkey virus, known as SV40, slipped through testing. As far as I've been able to determine, absolutely nothing happened to the many millions of people who received that virus along with their polio vaccination. And since that time, safety measures have unquestionably improved. But that isn't the whole story.

Anyone who sees the issue of biomedical research as only an ethical issue must do so by dismissing the biological realities. Those biological realities are staggering and far more pragmatically urgent. In 1986, *NOVA* rebroadcast a program on AIDS. That was the first time I heard it suggested that the AIDS virus was contracted into people via a chance bite or scratch from the African green monkey. It is that monkey who carries an AIDS-like virus, HTLV 4. The virus was believed to have then *mutated*, inside the human blood stream, to the AIDS virus, HIV.

Also in 1986, a report entitled Alternatives to Animal Research, Testing, and Education issued by the Congress of the United States' Office of Technology Assessment, stated that the African green monkey is used for corneal transplants into people. (As a nonbiologist, I dare not comment.)

The question of whether or not biomedical research is justified ethically remains a separate issue. Many picture such laboratories as dens of horror. There is, however, no question that the majority of biomedical laboratories are tidy, carefully managed establishments. The overwhelming majority of animals within them—with or without white jackets—are reasonably content. ("A rat is a pig is a dog is a boy"—is a biomedical researcher.) Scientists know perfectly well that neglected or mistreated animals are under stress. And stressed animals do not give reliable test results.

In one instance, the biomedical researchers chuckled grimly that various animal rights groups could only find two examples of cruelty in U.S. laboratories. But cruelty is cruelty, even if the examples are few. Biomedical research people are going to have to clean their own house or others *will*.

There is also the added reality that a number of leading scientists have concluded that much biomedical research is redundant—and it hurt them to acknowledge it. The notion that progress is being made is, unfortunately, only true occasionally—by no means is progress being made in any noticeable way in far too many areas. No matter how coolly the evidence is examined, an appalling amount of such research is just plain cruel with no redeeming features whatever.

From a special interest view or not, humanity must face up to the cruelty issue. Cruelty of any kind is no longer acceptable, anywhere, anytime, under any circumstance. Though the basic idea that cruelty is wrong has been around since Thomas Aquinas' time (1225–74), it's only recently that the concept itself has been carefully examined. Thomas Aquinas believed[17] that there was something deeply wrong with people who practiced cruelty.

According to James Serpell, in his book, *In the Company of Animals,*

> During interviews a significantly higher proportion of psychopaths reported inflicting cruelty on animals during childhood than a matching control group of convicted felons.

Because of our recent increased awareness of the inherent characteristics of cruelty, people are newly alert to signs of cruelty in all circumstances.

The practice of cruelty is fraught with hidden costs. I've always been impressed at the clarity of the memories surrounding any incidence of cruelty in both the witnesses and the direct participants.

As a single anecdotal example: I talked to one elderly man whose unit had a camp mascot in World War I. When it was time to be shipped home, the men agreed to dispose of the dog by taking him out to sea in a launch and throwing him into the water. The dog, trusting them completely, ran gleefully up and down the boat, delighted to be with the six men. When the boat was a mile or so out, the men tossed the dog into the sea and sped away.

More than half a century later, this man vividly remembered all the details of that day: the exact spot they launched the boat (a pebbly beach); the details of the weather (it was cloudy and chilly); the precise way the dog cocked his head when spoken to; how all the men involved avoided mentioning the dog at reunions; the nice things that this man has done for every dog he has bumped into since that time. It was a relatively small event in his life, but it had a tremendous impact. As he told me, "At the time, it seemed an easy solution to an awkward problem." He hadn't expected to feel so lousy about it for so many decades.

Of course, there are those who enjoy cruelty for cruelty's sake. Public opinion is against cruelty, and public opinion can move the proverbial mountain. (If I enjoy being cruel to other animals, and feel I must conceal this enjoyment from public scrutiny, I'm forced to admit that, no matter how great the pleasure, people would reject me if they knew I enjoyed it. Unfortunately, it's likely that my tendencies show themselves in other symptoms and will eventually give me away.)

One way or another, cruelty must diminish. Until biomedical clinics can open their doors to the public freely, people will believe the worst of them. As long as the public is aware of possible cruelty behind any closed door, efforts at stopping it will continue. Not only is it perceived as paving the way to further extant behavior, but also "...cruelty is vicious in its own right..." as Mary Midgley points out.

Perhaps the most difficult belief to overcome is that a great deal of medical research involving other animals is necessary. An appalling amount of it isn't in the least necessary. This is not my personal opinion: a growing number of major scientists from around the world have taken this position.

And those people who oppose cruelty in formal institutions are hardly powerless to change this sort of established precedent. One lone man (Henry Spira) created a nationwide campaign in the United States against a major cosmetic company (Revlon) to change their product safety tests for mascara. The details aren't important. That one individual changed the precedent of an enormous establishment—*that is important.*

The institutionalized forms of cruelty (even if only occurring occasionally) have another problem attached to them: they encourage the use and abuse of other animals in the private sector of society. As long as cruelty is sanctioned in one area, it is more acceptable in other areas.

Probably the single greatest obstacle in the way of a worldwide halt to the practice of cruelty is that there is no universal definition of the word. At first glance, that may seem a small obstacle: each person recognizes cruelty (or kindness) when he sees it. But a definition is sorely lacking, and cruelty is not all that easy to define.

There are at least hundreds of thousands of people in settled areas around the globe who strongly believe it's the ultimate kindness to allow their dog to "live his (or her) own life," left free to run in cities, suburbs, and rural communities. Anything short of complete freedom for the family dog is perceived as cruelty to such people.

Yet an equal number of people feel just as strongly that to allow the family dog to run free is the ultimate in cruelty. Free-roaming dogs are subject to being hit by cars, torn apart by other dogs, attacked by gangs of children, and, perhaps shot at in some places. There is always the possibility of a dog-catcher or game warden becoming involved.

Equally confusing is whether or not it's cruel to use a crop or whip on a horse in a race. Some folks feel strongly that it's cruel. Others feel just as strongly that a horse that will not run without such an incentive (a crop) will wind up as a can of dog food or a jar of glue.

The issue of cruelty is simply not as clear as it needs to be before a workable definition can be clearly stated. One of the ultimate cruelties is to allow any animal to suffer through long-term medical treatment, whether saving his life is or is not a certain result. Many veterinarians openly discourage people who try to do this with a pet cat or dog that has been badly injured. (Human doctors, on the other hand, often encourage such efforts.) Many people disagree with this view; they feel that letting any animal die or electing to kill even a severely injured animal is the ultimate cruelty.

Nearly everyone can easily describe an incidence during which they've seen an animal killed by kindness. And just about everyone understands that cutting a puppy's tail off a little at a time is anything but kind.

There are many other people who understand and tacitly agree with mercy killing, but do not agree if the other victim is a human animal. Yet one animal organization upset a goodly number of people by referring to cats and dogs as *cruelly murdered* for human consumption in a part of the world where cats and dogs *and* starving people were in close proximity. Unquestionably, there are killing methods that would universally be declared cruel, but the word *murdered* tossed in there made me doubt the entire business as too emotionally loaded.

In many areas, wording appears to be the whole point. Lots of people aren't in the least upset with the idea of some kinds of animals being slaughtered for human consumption. The United States' government, wise to the wording problem, now refers to the hunting and killing of other animals as *harvesting*. Terminology is everything.

Hunting (or harvesting) wild animals in the United States holds a unique place in the issue of animal rights and wrongs. In the United States today, few hunters actually hunt for meat; it's much more common to hunt for sport. I have personal problems with the idea that causing the death of another can be fun.

But fair is fair. Hunters, as a united group, have done more toward saving habitat than many of the other groups who express the saving of habitat as their main goal. Collectively, hunters have insisted on conservation in as many areas as possible for generations—and have done so since long before anyone else was interested.

The entire issue of hunting is no longer black and white. Some of the major conservation groups in the United States (most notably *Defenders of Wildlife*) have pointed out that hunters have enjoyed far more of the federal budget for the management of game animals, far more of the budget than bird-watchers and photographers have obtained. And today, the latter groups outnumber American hunters about two to one. Understandably, it's irritating to birders and photographers to know the bulk of their tax money is being spent on game animals only. Because of this, the issue is in constant flux.

As the basic dichotomy of the hunting issue makes clear, it isn't always easy to discern who is doing other animals good and who is not. My biggest personal problem in this area is with the animal rights groups; I've never been convinced they do more good than harm. The earliest animal rights organizations seemed determined to achieve an upper middle-class lifestyle for every giraffe, every dog, and every cow. The problem with this, as I see it, is that the world is quite full of people, many of whom are doing their level best, to feed and house themselves and their families. Such people feel downright betrayed when they see a giraffe or a dog or a cow who, at the whim of another person, is receiving more rights, more concern, and more guarantees in life than many humans can ever hope to attain.

When an impoverished person can see photos of an individual of a completely different species who wants for nothing; when elderly people must clean up trash spread by a neighbor's well-fed pets; when a child is attacked by a free-roaming pet out on a lark, a pet who has more rights than that child may ever hope to attain—something is out of whack in that society. Equal rights for animals too often means *only* the rights of *other* animals. Beyond question, rights are currently imbalanced in favor of the human animal: of necessity, those human rights must be real and absolute before there will be any chance of many other animals having rights—equal or otherwise. After all, rights originate in the human community.

And that brings me back to the earlier question of a firm definition for *cruelty*. Such a definition must be arrived at for all animals, even the human ones. In the meantime, rights will unquestionably remain slanted toward the one species.

The more general situation of rights has been steadily improving all around the world. Some years back, there was a vocal fuss (in the United States) on the subject of cruelty involving a male elk (wapiti) who was found dead in the state of Idaho wearing a radio telemetry collar. In the very best interests of elk everywhere, his herd was being monitored through the winter months in a direct effort to improve winter conditions for the entire herd. The one elk (with the collar) served as the monitor.

Nearly everyone who heard about this elk's death was upset. The field monitors, with the noblest of intentions, had affixed the radio telemetry collar to the base of this animal's neck. The collar weighed twenty-five pounds. And each time that elk lowered his head to graze or drink, that twenty-five pound collar slid the full length of his elegant neck and slammed him in the head. He died of starvation.

By this time, nearly everyone has seen at least a few films of herds of sentient beings being chased across craggy cliffs, or across plains, or through jungles, or over savannahs or endless other sorts of terrain. (I've wondered, at times, why there aren't more midair collisions between helicopters pursuing the same group of animals.) In some of these films, the animals being pursued are fleeing with their mouths open, gasping for air. And eventually, some (or all) of them are shot with dart guns, captured briefly for the collection of data—data on things such as coat condition, tooth size, age, general health, parasite infestation, and on and on. Then the individual who was darted is given an antidote and left alone to puzzle out exactly what happened back there in that confused dream. ("Wasn't I in the clutches of a strange-smelling predator just a minute ago? What happened?")

I witnessed one film where a man in superb physical condition (the well-known Jim Fowler) ran down a buffalo in 110° heat to practice tail-throwing. Simply put, Mr. Fowler had a choice of whether to run in that murderous temperature; the buffalo had no choice in the matter. He was forced to flee Mr. Fowler. (The buffalo was caught and thrown.)

The motives of those involved in the radio collar/elk incident were unimpeachable. As a direct result of that elk's death, radio telemetry collars have been designed until they are now often so small and light, the animals wearing them hardly notice.

In moving animals by herding them with helicopters or running them down (as Mr. Fowler did) or darting them to collect data—all of these incidents have something important in common. Beyond doubt, occasional individuals have died from the stresses imposed on them by such maneuvering; but they didn't die in vain.

As a direct result of these and many similar projects, public interest in other animals is at an all-time high. People's concern about other beings' stress, suffering—their very survival—is relatively new in the human world, and the concern itself was unquestionably brought on by all of the incidents of the sort mentioned above. It has resulted in the consideration of animal's needs under any and all circumstances. And that is a right that they simply didn't have a short time ago. Because people today see so many other kinds of animals being taken seriously by so many other people, the right of all to be taken into consideration has become absolute.

On a larger time scale, it was only yesterday that this right didn't exist. In that yesterday, it was commonly agreed that animals had no more ability to feel pain or pleasure than sticks and stones. As other animals became more familiar to more people, it was recognized that suffering is suffering, no matter who the sufferer is. At first the change in people was subtle; the bottom line didn't change much. It wasn't uncommon for some people to say, "Yeah? So they suffer. So what? If they suffer, who cares?"

It has only been as human conditions improved that people began to show concern over suffering. As a result, people have brought into being the simple right to be considered, the simple right to live. As we became more involved with other species, our awareness of them has in turn heightened our perception that our mutual survival needs—food, air, water, and living space—are pretty much the same from one to another. The quality of human life is heavily dependent on the quality of life in general—for all animals.

While the right of other animals to simply exist is relatively new, awareness of the extinctions of entire species is not new. As a human concern, extinction actually has a history. (It is only our efforts to halt extinctions—recognizing our part in causing them—that's relatively new.)

As a part of history, extinctions go back a long way. It was dutifully recorded that the last dodo died between 1681 and 1691. The only living great auk was killed on June 4th, 1844. The last quagga died in the Berlin Zoo in 1875. These individual incidents (and a number of similar incidents) were dutifully recorded at the time. Even if no one was concerned back then, the seeds of concern were developing. (After all, the losses were recorded.)

Recent recorded extinctions are much more rapid, more common, far more numerous. There is little doubt that, today, people are aware of the extinction issue. Both the issue and the concern are global in scope.

There is a definite correlation between human awareness of the sheer number of other species—both plant and animal—and the *rate* of extinctions. Had the dog gone extinct in the late 1600s, as did the dodo, everyone would have noticed. It would have been duly recorded, as was the disappearance of the dodo. But if the snail darter had gone extinct at the same time, *no one would have noticed*.

There are vital points about this that must be noted. First, anthropologists to zoologists—and the vast number of sciences between those two alphabetical extremes—have made Herculean efforts to make people aware of how dangerous it would be to lose still another living type of plant or animal. And clearly, the overview is quite as murky as these scientists have painted it. But the situation is not as hopeless as I've heard expressed too often. ("We might as well kill off the rest of the chimpanzees through biomedical research as a benefit to humankind; they're going extinct anyway.") Humankind isn't in any position to ease back on global efforts to stop the extinction of any other life form; we probably won't be for a long time to come.

But there's a backlash to this: while our awareness of the incredible number of other life types has been increasing steadily since the late 1600s, it also becomes clear that the complete annihilation of other types of life has been going on since that time. The rate of extinctions has been escalating since then. And that's the grim side.

The flip side is that millions of people around the world are working hard to halt these rapid losses of species (again, of plants and animals).

One of the many side effects of this has been the continuous growth of public awareness of the importance of the issue. In nearly any nation on Earth, I can ask the smallest schoolchild or the most aged adult why other animals are important to people and, in response, get a sensible answer, no matter which extreme I ask. Whether I ask the question in a hut or a town hall, I'll get a comprehensive answer. Awareness of the importance of this issue is *everywhere*.

In the process of granting rights to other species, people also lost a right. The fact that we never really had the right to pick and choose what other species would live or die is irrelevant. We believed we could do that. At times, millions, sometimes billions, of dollars have been spent in serious human attempts to eradicate entire species that, for one reason or another, were considered undesirable. And at some of those times, people have focused wealth, power, and massive resources in deadly attempts to eliminate the flea, the tick, the mosquito, the rat, the coyote—lots of species in different times and in different places.

Even before we realized we didn't actually have the right to perform such selective exterminations, people had to face the realization that selective exterminations could not be accomplished, no matter how much money or effort we expended.

The outstanding feature of the list above (and nearly any addition to it that one could make) is that *not a single species has been removed*. Not one. I was impressed to realize that people have never succeeded in deliberately eradicating another species of animal, though human efforts to do so, in some cases, have been formidable. Conversely, we have unintentionally eradicated far too many for comfort.

The passenger pigeon is a perfect example of an unintentional eradication. That animal was wiped out in a single human lifetime, though it was one of the most numerous birds on the North American continent. In 1886, eighteen years after Henry Bergh founded the American Society for the Prevention of Cruelty to Animals (ASPCA), a single flock of passenger pigeons passed over a small town in Canada, obscuring the sun for fourteen hours. Flocks of such size were common at the time. It was believed impossible to damage such a populous animal. (People have learned a great deal since then.)

In spite of their incredible numbers, the passenger pigeon never stood a chance. Young and old were killed indiscriminately in their nesting sites. Because many young birds never got the chance to reproduce, their numbers diminished with astounding rapidity. Single mass kills were frequently estimated to be as high as a million birds. Embarrassingly, our recent ancestors devastated this species in such astronomical numbers for hog food.

By the time it was realized that the passenger pigeon was in serious trouble, it was too late. Since many people had seen the large flocks so recently, it was impossible for them to believe what was happening. ("There's other huge flocks around; I've seen them myself.") The last passenger pigeon died in the Cincinnati Zoological Park on September 1, 1914. According to biologist Paul Ehrlich, her name was Martha.

Not able to find any evidence that we'd ever successfully eradicated a species, I began to question other people about it. And I learned that many people believed we'd successfully eradicated the Rocky Mountain locust. But that's not so. While there was an undeniable war between the United States and that animal, the United States didn't win it. Then again, neither did the locust.

In 1874, just before the beginning of the passenger pigeon tragedy, the Rocky Mountain locust made itself the scourge of the western plain states in vast migrations. While earlier records proved that the locusts had appeared many times in the past, no previous mass movements had the impact of 1874.

Like the passenger pigeon, locust numbers were estimated by measuring time and distance.[18] A single town reported one swarm estimated to be 300 miles wide, half a mile thick, and long enough to pass overhead for six hours.

The year 1874 was more than a quarter of a century before the Wright brothers lifted off the ground. And even if people had been able to fly well enough to get above any such swarm, what sort of insecticide or pesticide (neither of which were yet in common use) could have been used on that vast number of insects? Which one would have worked without killing every living thing under the insects as well?

Wherever those locusts landed, they ate everything edible, using *edible* in the broadest possible way. Reports of those locusts eating everything green and growing on hundreds of acres were common. And it wasn't just one swarm; they came swarm after swarm. Locusts were often found to be six inches deep on the ground. Men reported having to tie their trouser legs shut to keep the locusts off their bodies. (No one reported how the women managed and, since the plumbing of 1874 was outdoors, the women had to be managing.)

Since the worst of the trouble began nine years after the end of the Civil War, the U.S. Army was called upon to distribute food collected for the remaining settlers by a variety of organizations. Fully half the original settlers fled back east in desperation. Those settlers who remained were often forced to eat the locusts, just to survive. They ate locusts indirectly, by consuming poultry that had fattened on locusts; they ate locusts directly as well, in soups, cakes, and simply baked plain.

The immediate terror was bad enough but it paled considerably when everyone realized that all the eggs laid by the passing hordes were going to hatch. In an effort to reduce the likely resulting horror of that, all of the states involved awarded bounties for egg collection and, later, for hatchlings.

The newly hatched locusts did not become (and would never become) the terror their parents had been. In fact, the new hatchlings didn't even look like their parents. They looked like grasshoppers, just plain grasshoppers.

But the plague of locusts continued in the western plains of the United States until 1878, though they never again created the devastation of 1874. The very last appearance of a small migratory swarm of the Rocky Mountain locust was seen in Manitoba in 1902. And then they vanished.

I can just imagine the tension of those settlers, dreading the next swarm, and at the same time, trying to prepare for it. Some believed the swarms were caused by the imbalance created by loss of the great bison herds, but nobody really knew. Everybody had to wait for the proverbial other shoe to fall.

When it did, it probably seemed somewhat anticlimactic. According to Howard Ensign Evans, an award-winning entomologist:

> Shortly before the turn of the century, prospectors in the high country of Montana...discovered a glacier that was unusual in having horizontal layers of a dark substance. Close examination showed that these layers consisted mostly of Rocky Mountain locusts...each layer in the glacier contained uncountable millions of locusts, and each probably represented one or more great migratory swarms...

No Rocky Mountain locusts exist today, as far as is known. They are an extinct species, but not because people were able to eradicate them. It seems that their extinction came about through, of all things, the vagaries of wind currents. There are still plagues of locusts all around the world, including on the North American continent. But the Rocky Mountain locust is never among them.

Only now is it understood that migratory locusts of all kinds are just plain grasshoppers that become physically altered to a migratory phase when they become overpopulated. When grasshoppers become too densely populated, their young hatch with shorter hind legs, longer wings, darker coloration, and other minor differences—the migratory phase of the plain old grasshopper. On a global scale, vast efforts are made to control grasshopper numbers, just since this basic understanding of locust plagues came about. Though such efforts are continuous, they are not entirely successful.

The major Rocky Mountain locust plagues of the southwestern United States were rare in a number of ways; overall, the extinction of that species, no matter how thoroughly people hoped for it, was a natural accident. Today, by far, the majority of such accidental extinctions can be directly attributed to pollution problems, alteration of existing ecosystems, loss of habitat, endless more problems. But genuinely accidental extinctions, as in the case of the Rocky Mountain locust, remain rare.

The bottom line is that people have never actually had the skill or the right to deliberately and selectively destroy an individual form of life. People do not have that capacity—we can destroy *all* life or we can save *all* life. But wherever humans have tried to play selective games with that reality, the result has been incredible.

In our magnificent grand scheme to control a single species (the mosquito) in the DDT fiasco, we damn near succeeded beyond our worst nightmares. We came very close to unraveling the entire food chain of the biosphere. Because people recognized the impending disaster in its early stages, we were able to save the birds, the food chain, and quite definitely humanity itself.

But it was close.

We are just beginning to find verifiable proof of the slow recovery of all the systems that were damaged by just that one bumbling effort to selectively eliminate a single sort of being. For an awfully long time, it was uncertain if we'd ever recover from those damages.

It was a close enough call to scare some sense into most of us. Close enough to provide a tiny peek at what might happen if we actually won the war against nature, the war we'd felt so free to fight enthusiastically just a short time before. Close enough to create the vision necessary for people to perceive that other animals *must have the absolute right to be considered*, the absolute right to exist—maybe not everywhere, but nearly everywhere. With truly remarkable speed in time, the tacit understanding that all creatures have an absolute right to simply *be* has spread around the planet. It has become a universal principle among people, and extends from the soggiest swamp to the deepest jungles, from remote mountains to the heart of urban dwellings.

It was only when I had gotten this far that I began to understand the overview of where people really do (and really don't) fit into the kingdom of animals.

CHAPTER 10
DIVISIONS: REAL & IMAGINED

*I*t's only been a short time that the question of what separates
people from the rest of the animals has been around. For most
of human history, people thought of themselves as just a part
of the animal kingdom and nobody worried about it.

When early people looked around at the rest of the animals,
they decided there were only two basic kinds: predators and prey.
Though they had always eaten some vegetation, early people also ate
quite a bit of meat; predators ate meat. People obtained meat by
killing other animals; predators obtained meat by killing other animals. Voila! People are carnivores! People could kill and eat instead
of being killed and eaten.

A curious side effect of this early belief may have played a role
in our choices of pets. To be sure, there are herbivorous pets, omnivorous pets, and even insectivorous pets. But the vast majority of pets
around the world are carnivores. This may be more significant than
we've realized; it may also explain something unexpected in basic
human nature.

As various branches of science learn more about other animals,
it's been noted that carnivorous animals *play* more than most animals

in other categories. I've even heard one biologist remark that this tendency to play may be directly related to the form of protein consumed.

Herbivores also play, without question. But they play most frequently when young and still consuming animal protein, via their dam's milk. Adult herbivores are more inclined toward aggressive behavior. Carnivores, on the other hand, play throughout adulthood, as many of us do. In spite of that implication, people are not true carnivores. Humans are omnivores, like blue jays, hogs, bears, and many other species.

But this truth escaped us for a long time. Eventually, our ancestors apparently decided it wasn't enough to separate themselves from just some of the other animals. After all, *carnivore* was still an animal classification. People wanted greater distinction. And a clear part of the desire to be separate was a desire to be elevated. Put another way, we people wanted to feel superior. And that began the greatest number of imagined distinctions evolving into false divisions.

One of these false distinctions, as was mentioned earlier, was man as a tool-using animal. Being the only tool-using animal separated us by a skill that was both dignified and superior. As a difference, it was eventually discovered to be more imagined than real. From Indian elephants that use leafy twigs as fly swatters to sea otters that use stones to break open oysters, tool-use is relatively common throughout the animal world. Tool-use is an imagined distinction when it comes to people.

One specialist distinction first noted long ago that never became commonly known is the thickness of the human skull. It is that distinction that helps archeologists to distinguish human skulls from other ape skulls at digs. Simply, the human skull must span a large brain area so thicker bone is necessary to protect the brain. Most animals have much thinner skulls than do people.

Like the above example, not all human differences are imagined, by any means. Most frequently though, such differences are small and simple peculiarities among the animals. One such oddity is *hand-edness*. According to *Newsweek* magazine, 90 percent of people tend to be right-handed, while most other animals ". . . seem to be more evenly divided between left- and right- 'pawedness.'"

People are full of such peculiarities, small differences that don't seem to mean much in any comparative way. For example, any tracks that are found in the correct size category are always considered to be human if they are in a straight line. The reason is because so few other beings walk in straight lines. Humans do. There are a few small rodents who do. The European fox and the North American wolf nearly do, but not quite. This is another real difference between people and other animals, even if few of us notice it.

Jacob Bronowski (1908–74) proposed that one of the major differences between people and the rest of their kingdom is that we people tend to alter the existing environment to suit our own needs. And while the beaver also tends to alter his environment by controlling the flow of a stream in the way he builds a dam, and other animals alter their personal environment by digging dens or building nests, it is people who alter their environment the most. We try to alter every phase of our environment. We build a shelter, heat it, light it, and bring in water via a controlled system. People may even live completely within it if the season dictates. Other beings don't tend to make such thorough alterations. (Again, this is a difference of degree, but not of kind.)

Other beings are more inclined to adjust to the existing environment, even for seasonal changes. On a short-term basis, many mammals (such as cats, dogs, bison, and bobcats) grow longer/shorter hair for the winter/summer weather, gain or reduce body fats (like bears) as seasonal adjustments. On a longer-term basis, some animals adjust their entire bodies to suit specific needs—like the giraffe's long neck to reach food less accessible to other animals. Still other animals alter their behavior by becoming nocturnal (like the owl) or diurnal (like the crow) to avoid their personal environmental pressures: daytime heat, intense predation, and competition for food.

People regularly make the same sort of changes, but for different reasons: fashion may be the reason behind longer/shorter hair; people may gain or lose weight independently of the weather or season; personal preference or a work shift may encourage a person to choose nocturnal or diurnal habits. Generally, people are incapable of making such physical changes in reaction to the environment as other beings do.

But in small ways, that are just beginning to be noted, people do indeed have biological reactions to environmental pressures, just like the rest of the animals. A doctor once told me he had noticed that people's blood tended to thin in warmer weather. People eat smaller, lighter meals in summer. And in winter, there is recent evidence that quite a lot of people suffer from seasonal depression, just because the natural light available is so drastically reduced in that season. There's probably lots of similar changes going on in the human world all the time. (I was unable to find any documentation for the human tendency to be restless in spring, but we all feel it.) For so long, it was assumed there were no such changes so no one looked for them.

As time marches on, more similarities than differences have been surfacing between people and other animals. By tradition, we've briefly focused on this or that difference, much as the hummingbird focuses totally on one flower at a time. "This is the important answer/flower that supplies what I want." And that focus remains complete until the answer/flower proves empty. Then we focus on the next one just as completely.

For as long as I've been interested in what does (and does not) separate people from other beings, I've been alert to all the divisions I've heard announced. And, sooner or later, nearly every specialty has announced its version. I've usually ended up feeling frustrated by most of these answers, as I realized early that false distinctions tended to shadow the real ones, keeping the actual differences buried under the rubble of imagined ones.

As only one example, an education specialist insisted that *only humans are capable of learning to listen.* Yet when a deer stops grazing and raises her head to look around in response to a sound, is she listening? Is her listening necessarily of a wholly different quality than my own? Maybe, but I seriously doubt it. Such differences crop up with monotonous regularity.

Such divisions seem to stem from a deep belief that all other beings are somehow less in every way. Dr. John C. Lilly spoke to this human tendency: "In our present state of ignorance only a human being is said to have an intellectual capacity."

Like others, Dr. Lilly focused his attention not on what separates people from other beings, but instead on what should separate still other beings by the same reasoning that currently separates people.

Along the same theme, Jane Goodall wrote at length in the *New York Times* about her concerns regarding the way chimpanzees are treated in biomedical research: "Chimpanzees have given me so much in my life. The least I can do is speak out for the hundreds of chimps who, right now, sit hunched, miserable and without hope, staring out with dead eyes from their metal prisons." The theme of the article is that chimpanzees are intelligent, emotional animals that deserve better treatment.

Ivan T. Sanderson wrote eloquently in favor of separating gorillas from our general attitudes about other animals: [The gorilla is not] "...the ravening ogre he has been depicted but just a great big, easily scared vegetarian, desiring nothing more than to be left alone in his forest fastness to raise his solemn, quiet little kids, and be allowed the occasional privilege of marauding a human banana plantation."

These people, and many others like them, have demonstrated deep understanding of a particular species. Their comments suggest that these other beings be treated as separately as we treat ourselves from the way we treat other animals in general. Conversely, such statements make it abundantly clear how badly other beings can be (and often are) treated by people. These heartfelt comments fill me with compassion for dolphins, chimpanzees, and gorillas.

No one speaks for the others who may not be quiet or intelligent or gentle or friendly or even likeable. And they, too, deserve better treatment from people. While I admire the dolphin, the chimpanzee, and the gorilla, I also admire other animals, and some of them aren't so easily endeared to people. In all cases, other animals are far more like people than different from people. Singling out just a few kinds for better treatment implies that there exists a justice in the treatment of the other members of our kingdom. It also begs the question of what those differences might be based on: quietness, intelligence, gentility, friendliness, or simple likability as only some examples.

Each time it is successfully argued that another species be *includ*ed in our hierarchy and *ex*cluded from the rest of the animals, someone can come along and argue for yet another species. And that brings me back to the original question: What is the distinction between people and other animals?

As individuals, we draw our own conclusions based on personal experience, then doubt them. I find it difficult to achieve genuine objectivity in a subject I'm so much a part of. I've found it vital to disconnect myself, to try to look at the overview.

In doing so, one of the first things I noticed was that many species are better adjusted to life on this planet than people are. As an example, whales have developed an existence that fits into the existing ecology. There has been no need (real or imagined) for whales to damage or destroy their environment. They live lives based on sound ecology by functioning within the existing biological system.

Are people *ex*cluded from that kingdom because we damage the environment? Or is exclusion based solely on our desire to be excluded? Are we as alien from the rest of the animals as we think we are? The way the majority (of other animals) treat us, I'd be inclined to think so. Most animals treat people as wholly other beings—which of course, we are.

The point that stood out most to me about the incident between Mr. Gibbs and the bear was that the bear was so tolerant of Mr. Gibbs for so long. Bears are hardly noted for their patience, yet she allowed Mr. Gibbs to follow her and her three cubs for quite a while before she stopped him. Mr. Gibbs, and the many other people who have found themselves confronted by a powerful animal, clearly believed in that ultimate power of human superiority; after all, he bet his life on it. Since no one was present at that confrontation, I can only guess what actually transpired. It seems obvious to me that Mr. Gibbs was so thoroughly convinced of his own safety that he was able to keep that bear at bay just because his conviction was so great. It must be remembered, there was more than one mind involved in that incident. It seems reasonable that the bear was intimidated by Mr. Gibbs' boldness.

For many people, the conviction of human superiority is adopted quite unconsciously in childhood and is never really thought about again. For some, that assumption of superiority is believed to apply in all circumstances, under all conditions, and is believed to be understood by all other animals, including she-bears.

For myself, I was inclined to slide into this (il)logic from several movies I saw as a child. A lone cowboy, standing out in the middle of nowhere, faced a herd of stampeding cattle. The cloud of dust and the thundering of the cattle's hooves pass and there he was! He was standing there in the settling dust, looking amazed, and gratified that the cattle recognized his ultimate superiority, therefore they were careful not to injure him.

Eventually, I got to wondering what the ungulates (hoofed mammals) were thinking about while thundering down on this lone cowboy. Granted, they probably weren't thinking much, but they had eyes and they could certainly see him standing out there.

It was years before I came up with a way to see the big scenes in those movies differently. Riding horses in winter with a group of friends, we all had a hard time trying to get the horses to leave a cleared road surface, to take to the deep snow; the horses simply didn't want to make that transition. But on the return trip we came out in a different place on the same road, paved, and the horses practically rioted. They seemed convinced that the bare pavement was ice. Finally, one of the riders dismounted and led his skittish sorrel mare out onto the "ice" of bare pavement. Then, and only then, would the other horses risk it.

By now, I've seen plenty of such incidents, enough to have come to understand that ungulates live by their feet. For them, being able to walk is equated with being able to live. Just about all of them are afraid to step on (or in) something that might cripple them and force them into the solitary existence of not being able to keep up with the herd. Lameness often means death to an ungulate. Every moment of his (or her) being is ultimately dependent on those feet, from being able to graze to being a functional part of a social group. (I don't think there is a precise equivalent loss for humankind.)

So what about that lone cowboy standing out there in the middle of nowhere? Without question, some such incidents end with one very dead, very flat cowboy who looks no better than any other road kill. But in those incidents where the stampeding cattle don't injure him, it would seem that many people conclude the cowboy's survival is directly related to his innate superiority, and that the cattle (or whatever animals) recognized that superiority. Since people never see any other sort of being standing in the way of a stampeding herd of animals, there is no easy comparison. (In actual fact, other animals would be more careful than to get caught in such a situation.) But if I could get another being to stand out there, in the middle of nowhere, as a stampeding herd approached, I'm quite confident the herd would part just as readily (if possible) around any other animal standing there. The final scene would be, say a gorilla standing in the settling dust with a look of amazement on his face, pretty similar to the cowboy's. The herd would avoid the gorilla just as carefully and for the same reasons: strangeness and fear of damaging the all-important hoofs.

The theory of other animals perceiving humans as superior may be subconscious in us all, and I do believe it's there in most people. Though people may fail to personally notice it, most animals instinctively move away from alien beings, quickly and quietly in most instances. This seems to be equally true, no matter which animal is stronger or weaker. Familiarity has a lot more to do with this evasion than strength—real or imagined.

It may very well be this familiarity that is causing many of the problems between alligators and people in the state of Florida these days. About six weeks after the alligator killed the young child (as was noted earlier), there was another direct confrontation between an alligator and a person. In this second incident, the man involved (Joseph Hays) was twenty-two and, wise to the ways of alligators, he telephoned for help before he approached the injured animal lying beside the road in the dark, having been hit by a car. That small detail of calling before approaching the animal saved that man's life, beyond question.

After help arrived, Joseph Hays approached the ten-foot-long reptile. It swung around with astonishing speed and caught Mr. Hays' arm. I'm certain they were all remembering the incident with the child just weeks before. The three law enforcement officers shot into the alligator ten times before it would let go of Mr. Hays. Later, one of the officers referred to this incident as the most terrifying thing he'd ever witnessed. Before it was over, however, it took eight more bullets (for a total of eighteen) to kill the now mortally wounded alligator.

It's important to note that the alligator involved here had already been hit by a passing car, that it was already wounded before the confrontation with Mr. Hays. Even under those conditions, it was still able to move with a speed that clearly caught all the people present off-guard; it still had the strength to hold onto Mr. Hays through many gunshot wounds. The fantastic strength and speed of other beings catches a lot of people unprepared for a mighty adversary—an adversary the people had expected to be inferior in all ways.

No matter how it's analyzed, confrontations between people and many other sentient beings that are unknown to each other do not have predictable results. Such confrontations always involve more than one mind, more than a single way of perceiving the meeting, more than one idea of what the outcome should be. And it's that reality that catches too many people unawares. There is no automatic division between people and other animals through innate superiority. While quite a lot of people seem to expect such a division, it really isn't there.

Of course, there are real divisions, but they are rarely overall divisions. While human senses are relatively well-defined, it's commonly understood that many other animals have superior senses: eagles (and most birds) can see better and farther than can people, dogs can hear better and in a much wider range of sound than can people. Further, it's long been established that some dogs (mainly bloodhounds) can clearly distinguish between the scent trails of identical twins.

Directly compared to the rest of the animal kingdom, people's senses are adequate, and that's about all. In some of the senses, people are practically at the bottom of the list of perception. Like the rest of the primates, people have a sense of smell that is rated poor to fair.

Other senses, like taste and touch, seem less easily measured, more subjective in degree. The sense of taste is closely tied to the sense of smell in people. Yet the eagle (who can see farther than people) may have no sense of smell whatsoever.

Science vacillates hotly from one extreme to the other on the subject of whether birds can smell or not. For quite a long time, it was believed that the kiwi of New Zealand was the only bird who had a sense of smell. Yet many sorts of birds are known to reject certain kinds of caterpillars because of bad taste. And that implies that the senses of taste and smell may not be connected in birds, or that birds do have a sense of smell.[19]

The sense of touch strikes me as the ultimate subjective sense. I've never touched any animal that didn't obviously notice my touch immediately. I sort of backed into understanding of the subjectiveness of touch long ago, when I began wondering why cats and dogs enjoy being stroked. While some of their enjoyment clearly has to do with reassurance, familiarity, security, proof that no harm is intended, both cats and dogs also seem to enjoy stroking for its own sake.

When I compared my own reactions to being touched, I noticed that if my arm was touched, the feeling seemed blunt. But if my hair was touched gently, the feeling was greatly intensified. That would seem to explain why furry cats and dogs get such pleasure from gentle stroking. There doesn't seem to be any reason to assume other beings have a greater or lesser sense of touch than people do.

Years ago, I spoke to an anatomist about the differences among animals, and he told me that there were essentially none among most mammals. "You open a pig or a person, there's vastly more similarities than differences. Hearts, lungs, stomachs, livers, intestines...they're all pretty much the same from one animal to the next." He paused and looked at me closely. "Don't misunderstand; there *are* differences. They're just small."

Many other human distinctions fall largely into the *may be* category; we may be different here or there. Human inclinations toward creativity in the varied forms of technology may very well be unique in the kingdom of animals. As of now, it looks as if they are; but often, to state such a thing conclusively shows our ignorance more than any provable fact.

Music, as one example, may or may not be what the song of the humpback whale is all about. It's possible that religion, philosophy, knowledge of the past, and wishes for the future may all be signs that indicate human. We have no idea, really. Such subjects are likely to remain dubious distinctions until we can actually communicate across the barrier of speciation. As they all have the potential of being no more than imagined divisions, I feel it would be unwise to rely on any of them. Any evidence currently assumed to prove anything in these areas most often is the evidence of absence: "Well, we have yet to notice anything we consider a sign of creativity, religion, philosophy, (etcetera) in any other animals, so those things may be exclusively human."

Personification—the tendency to humanize inanimate objects—may be unique to people. But, like many another seeming uniqueness, its exclusiveness may simply be the word we choose: *personify*. It may be that whales *whalify* things, gorillas *gorillify*, or what have you.

As long as the hunt continues for some small (but definite) distinction that belongs to humankind and only to humankind, we aren't likely to perceive the biggest difference between people and all the other members of the great kingdom of animals.

As the world becomes more crowded with people, it has become necessary for people to specialize in order to achieve anything of note. One result of this has been that it's become more difficult to perceive *Homo sapiens* as a whole. People become fragmented into isolated groups by nation or neighborhood, by profession or age group, by earning capacity, by marital status, by race, by becoming parents, by membership in clubs, by religious affiliation, and the list could certainly go on.

With all these divisions between people, still the greatest distinction of all has remained the same, unchanging throughout history; it hasn't even changed since before history was an idea.

In order for a distinction to be genuine, it must apply to any and all people, from the ancient past through today. And this distinction does. There is no other animal as versatile as *Homo sapiens*. There were none in the past and there aren't any others today.

Any person may score highly on intelligence tests—or not. *Any* person may be skilled with language—or not. *Any* person may have great dexterity—or not. *Any* person may be able to create art or a concept as basic as the wheel or a new way to fly—or not. There are at least millions of people with no creative tendencies at all, but *any one* of those millions might be creative or dexterous or intelligent or talented in an unexpected area. Every one of those millions—and the billions more like them—has vastly more potential in a wider variety of areas than any other animal, simply because they are a part of the most versatile species that has ever existed.

With only one recent technological exception (birth control), everywhere I've looked for a difference in *kind* I've only found a difference in *degree*, and our versatility is no exception. However, it is the most dramatic example.

If the great whales could come safely to land tomorrow, they might dazzle us with their intelligence, but they wouldn't have our freedom of movement: they would be unable to move about on land; they would be unable to shell peas or build a swimming pool or simply take a walk—all simple things that any human has the potential to do. Were chimpanzees given any section of the human world to do with as they chose, they might well learn to operate individual parts of it far better than whales could, but they haven't got anywhere near the human capacity to discern the need for a shoe, make it, or create the want for it in many of their fellow beings. But any person might be able to do any or all of those things.

It doesn't take away from human versatility in the slightest that people need technological aids to fly or make shoes or swimming pools. If anything, the human desire to create and use technological aids to overcome physical restrictions enhances the point. To be free of all restrictions is the ultimate human goal.

As a distinction, human versatility is magnificent. No less. It remains unchallenged by any other being, just as true today as it was in the past. Any person may create something like paint, then go on to paint a picture or even a house with it. Any person may create a song, then sing it and get others to sing it as well. Human versatility isn't limited by a person's specialty: a great surgeon may be an excellent swimmer; a great linguist may be nearly as articulate with a paintbrush; a great musician may also build beautiful houses.

This versatility holds equally true for people everywhere—from the most primitive to the ultimate sophisticate. A great teacher may teach spear-throwing or trigonometry; a great architect may practice hut construction or skyscraper construction. The human option for vastly diverse skills applies from village chiefs to leaders of large nations, and to children everywhere. The potential is present in every one of us. Wherever an individual is sufficiently exposed to possibilities, that individual may perform adequately or brilliantly in that area. Anything is possible where the human animal is concerned. Anything at all.

In this precise sense, human versatility becomes self-fulfilling: if a person can perceive possibilities, he (or she) is likely to be able to build on those possibilities. And this point cannot be overexaggerated. Please keep in mind that *it was less than sixty-six years between the first flight of the Wright brothers and a man walking on the moon!*

Few human accomplishments are anywhere near that easy to document. While it's possible to prove that concrete has been in existence since 193 B.C., how long was it before concrete was used to build the tallest structure in the world? And after all, there have been at least hundreds of tallest structures throughout time, each falling into the shadow of the next tallest structure.

Intelligence doesn't really have all that much to do with human versatility; it may help, but it isn't a necessary ingredient. In a single day, any human may use his (or her) hands for surgery or shucking ears of corn and a thousand casual and complex diversities between those extremes; there are also the other diversities peripherally involved in any such example: driving to and from the hospital, giving instructions, taking instructions, correctly evaluating which person to operate on or which ear of corn is ripe enough to shuck.

Intelligence, by itself, isn't really all that much to brag about. The animal kingdom is full of intelligence, to one degree or another. In fact, quite a lot of the other intelligences look as if they may roughly equal our own. Although many other animals may be intelligent, every other animal has become closely tied to specialization. People also tend to specialize, most definitely, but people specialize *as individuals—not as a species.*

Ultimately, that represents the finest, most clear, and definitely permanent distinction of humanity. Other animals specialize; people also specialize. Other animals cannot change their specialty, but people can. The giraffe must continue to browse the treetops. The lion must live on the savannahs where his yellowish coat is his best camouflage. The whales must remain in the sea. But humanity can collect the fruits from the treetops, live in a swamp or on a hilltop, swim with the whales and then return to a totally different atmosphere and go home. As far as changing specialty activities, people can always do that. Other beings simply cannot.

In spite of the reality that this is, by far, the greatest difference between people and other animals, it is yet again a difference in *degree*, not in *kind*. These days, we know far too much about the rest of our kingdom to be able to point to our heads or our thumbs and say, "This is it! This is our big difference!" Today, such a comment would be a simple declaration of our own ignorance. Many people now know that the most casually stated differences are probably more imagined than real. And many are aware that our ignorance remains vast, still much larger than our knowledge.

But when it comes to versatility, it's ours and ours alone. No other animal can accomplish the multitude of diverse tasks the average person accomplishes on an ordinary day. We have other differences, but no other difference between people and other animals is as large or more dramatic than this one.

CHAPTER II

PETS & THEIR PEOPLE

W hile modern people think of the domestication of animals and pet keeping as one and the same process, they are not. The function of the pet is quite different from that of most domestic animals.

Though many animals cross back and forth over the line between simple domestication and being pets, others do not. The dog is everywhere considered to be a domestic animal, for example. Yet dogs are regularly used in bewildering ways, from one extreme to another, all around the world. Dogs may be considered slaves for everything from sledge pulling to fighting; they are expected to fill varied roles from hunting with people for food to being the food themselves. Dogs also vary from being considered paragons of virtue to the ultimate example of filth, from instruments of home protection to members of the innermost family circle. What the dog's role is as a domestic animal depends precisely on the time and place the question arises; being a pet is only one of many possible roles.

On the other hand, any animal may be considered a pet. In the United States, the industry of exotic pets—animals that are far from domestication—has been steadily increasing. Lions, tigers, zebras, wolves—nearly any individual animal (of any species) can be included in the pet category.

As mentioned in an earlier chapter, carnivores are the most popular pets. Part of the reason for this is their tendency to be more playful than herbivores, but there are other reasons: carnivores tend toward tidiness in confined spaces (like homes); their food can be a simple dish on the floor. Herbivores, as an alternative, need large and bulky food sources like piles of hay, which are far less convenient.

The herbivore's bowel also tends to function on an automatic basis, with no conscious decision on the part of the animal. Carnivores have a good deal more control over such things and adapt to any reasonable schedule.

Carnivores also sleep a great deal more than many other sorts of animals, a great boon to human schedules. For these subtle reasons, carnivores will likely always outnumber other animals as pets.

In spite of the current popularity of exotic animals, domestic animals vastly outnumber all other animals in the pet category. Through the world, cats and dogs are still the majority of pets. The popularity of pet keeping has been on a steady rise all around the world. In the United States alone, in one eleven-year period, the number of cats kept as pets jumped 10 percent, and the number of dogs jumped more than four percent, in the same period.

It's not likely we'll ever know exactly when and/or where people began keeping other animals as pets. It probably began with the dog. According to Desmond Morris:

> It is probable that it began as a result of young puppies being brought into the tribal home base to be fattened as food. The value of these creatures as alert nocturnal watchdogs would have scored a mark in their favor at an early stage.

Since that long-ago time, pet keeping has remained a part of the human scene, though its popularity has fluctuated violently at times. By far, the most dangerous time to keep pets was between the years 1500 and 1700 in Europe, as any pet owner was subject to being tried as a witch.

There have been extraordinary ups and downs in keeping pets. Most of these have been recorded elsewhere. The more recent history of pet keeping is usually ignored.

For example, in the early 1950s, pet keeping took an odd turn in the United States: anyone who owned a pet was frequently subjected to an on-the-spot analysis by parlor psychologists who would eagerly explain why a pet owner kept pets. The most frequent reasons these analysts gave were: that pets served as substitute friends, as substitute lovers/spouses/children—always as a substitute for something. Such analysts were never deflected by those who kept pets and had friends, lovers or spouses (sometimes both) and children. In spite of a such social harassment, millions continued to keep pets.[20]

During the same time, European thought arrived at an entirely different conclusion. European professionals declared those without pets were people who were less capable of concerns beyond self. In spite of this social harassment, millions continued not to keep pets.

Oddly, the dichotomous positions of the United States and Europe never confronted one another. Instead, the situation took an unexpected turn. More and more scientists around the world began reporting on the rapidly declining quality of our mutual atmosphere. The end result of this has been an all-encompassing concern for animals in general. And once the public became concerned, the entire issue of pets and people paled in significance. Just about everyone could see that the concern for animal life was the more important issue. Where technological advance made it possible to quickly discover many *new* animals on the one hand, those new discoveries were dying out of existence as quickly on the other hand. As technology grew, so did the number of extinctions.

Technology made it possible for more people to know about more kinds of animals than ever before. As only one example, in the United States, there were 4 million television sets in 1950; by 1959, there were 52 million sets. And on those television screens, in the tradition of Aesop, Walt Disney became widely known by telling stories using other animals' lives to represent human problems in fable form. New methods of book distribution renewed interest in some old books: Anna Sewell's *Black Beauty* enjoyed a new burst of popularity. And *anthropomorphism*—attributing human motives and emotions to other beings—became all the rage.

Early in 1960, Dr. John C. Lilly pointed out that in reacting (and overreacting) to the new public attitude of anthropomorphism, many people began to practice *zoomorphism*, where no motive of any animal was considered to be much above the fundamental drives of reproduction, seeking food, or some form of self-gratification. Once zoomorphism became regularly used to describe other animals, it was only a short step to interpreting human motives in the same way. (This is still commonly practiced here and there.)

In the meantime, an entirely different scientific approach to pet/people relations cropped up. It's best described as "Don't interpret—just watch."

The information provided by this new approach was unexpected: no matter whether people have an anthropomorphic, zoomorphic, or any other sort of relationship with their pet, people who keep pets have distinct advantages over people who don't keep pets. When exacting studies of relations between pets and their people were run at the University of Pennsylvania, it was discovered that a pet owner's blood pressure dropped markedly in reaction to being presented with his (or her) pet. Additional studies along the same lines proved that those people who keep pets recuperate more quickly and more thoroughly from serious illness than people who do not have pets. This has all now been proven.

But there's a whole other side to this issue that is too often ignored. What is going on in such situations for the nonhuman animal? What benefits do the other animals in these experiments get?

In his *In The Company of Animals*, James Serpell reported "ethically disturbing" experiments that took place in the 1970s at Johns Hopkins Medical School, where dogs were given repeated painful electric shocks to monitor their heart rates and blood pressure. Rather predictably, the experimental dogs' heart rates doubled during electric shocks. And when a warning tone was added to the experiment to alert the dogs to the imminent shock, the dogs in the experiment reacted to the tone alone with noted anxiety. Someone came up with the idea of including a person in these experiments, to see what effect this addition might have on the dogs. According to Serpell, "When an experimenter sat with the dogs and petted them while the tone and shock were delivered, the usual marked increase in the heart rate was either eliminated or changed to a decrease in the heart rate."

It's pretty obvious from the above that while pets cause a decrease in the heart rate of people, the reverse is also true. Oddly, there's one subtle difference between the two perspectives: people need to be presented with their own pet before their heart rate changes. The dogs in the experiment, however, were simply presented with any person and their heart rate would slow.

Overall, there is an unexpected similarity here, specifically involving heart rates, both between pets and people, and people and other animals. But it is uncannily opposite to the normal situation where people tend to generalize about other animals: "I love dogs," or "I love cats." And in the same generalized way, other animals are usually quite specific in who they care about.

While dogs were the only animals used for this experiment, I'm confident that other pets would have had similar reactions under the same circumstances.

Human interpretations of what is/is not going on during the

meeting of two beings should at least be open to question. Recently, there have been considerable mentions of the cat's tendency to brush against the legs of people he knows in greeting. It's nearly always stated that the cat is scent-marking the person. But it's said in such a way that implies that is all that's going on. There's reason behind this assessment: many kinds of animals do similar marking to places that are familiar. But when I tried to look at this common activity a bit more objectively, I saw something unexpected.

When any person who knows horses walks up to a horse, whether that horse is strange or familiar, the person always touches the horse in several places, rubbing his hands on the neck or shoulder, usually talking all the while. Looked at from an objective step away, this common activity doesn't look a whit different than what any cat does to a favored person.

The reason the person who knows horses does this is to reassure the much larger animal that the person intends no harm. But it also serves to put the horse on notice as to where the much smaller being (the human) is located, so the horse doesn't swing around or step on the smaller person.

If we zoomorphize what the cat is doing, why not zoomorphize what the person is doing? Is the person scent-marking the horse? (Maybe.) Is the cat warning the much larger human not to step on the small cat? (Quite possibly.)

Since cats and dogs are by far the most common pets among people everywhere, the issue of preference should probably be mentioned. (Incidentally, I refer to cats first because C precedes D in the alphabet.) But I believe that most people who are seriously interested in animals simply don't have a preference between these two. Still, some people have adamant preferences.

While all animals commonly create some emotional response in people, cats most often create strong responses, pro or con. To the ancient Romans, the cat was a symbol of liberty; in Egypt, the cat was worshipped; during medieval times, some cats were executed as witches. Beyond a doubt, cats run the gamut of human opinion with no say whatever in the final outcome. Humans are rarely ambiguous about cats.

Out of curiosity, I've made it a point over the years to ask each cat hater I've bumped into why they hate cats. Eventually, I noticed that the same points are always mentioned: cats are considered fussy in all ways, and cat haters resent that. Cats are fussy about food (they will not eat just anything offered to them) and cats are fussy about who touches them and when.

Eventually, I recognized that I was just as fussy as cats: I will not eat just anything presented to me as food, and I certainly object to being handled by strangers (or even loved ones) at inappropriate times. I'd guess that's true of everyone. So the very traits some people don't like in cats are directly correlative to people. People like to live lives of dignity and independence, as tidily and as free as they can manage.

Briefly, cats can be very affectionate, cheerful, and are often quite obedient. Cats frequently exhibit high intelligence, beyond formal expectations of their capacities. Cats demonstrate loyalty to people they are fond of, loyalty to hearth and home, and often bravery beyond their realistic capacities. They are rarely submissive.

Dogs, as noted, have been domesticated far longer and more thoroughly than cats. Like people, dogs can eat nearly anything and survive quite nicely. Dogs can be affectionate, cheerful, and are often quite obedient. Dogs frequently exhibit a good deal of intelligence, often far beyond formal expectations of their capacities. Dogs demonstrate loyalty to people they are fond of, loyalty to hearth and home, and often bravery beyond their realistic capacities. They are frequently submissive.

Beyond these generalities, the two species are different. Entire libraries have been written about the finer qualities of cats. Equally large libraries have been filled with books celebrating dogs. One of the fundamental differences between these two species can be summed up briefly: dogs believe disobedience results in the wrath of the gods falling upon them; cats believe disobedience is the spice of life. And there are always a great number of individual exceptions to this thumbnail rule about the two, both in individuals of each kind and in different times and places in the same animals. Because of the commonness of both as integral parts of human society, there are a few points about their residency (in human society) that should be addressed.

Many people seem to hold unconscious expectations of both species that are unrealistic, even unfair at times; other people do not. Both extremes are everywhere apparent in human society: simply put, one side believes that in communicating with pets, everything is understood; the opposite extreme feels that nothing is understood.

Unquestionably, the truth falls somewhere in between. Cats and dogs understand a great deal more than we give them credit for, and viceversa. The first group, the *everything is understood* folks, frequently provide entirely too much freedom for their pets—more freedom than any human will ever get to have. The opposite extreme, the *nothing is understood* folks, freely abuse their pets through neglect.

Realistically, both cats and dogs are predacious animals and both are capable of inflicting serious bodily harm on people. As small as he is, the domestic cat can be pushed to where he can become exceedingly dangerous. Too many people take a cat's early warnings lightly, believing a cat to be a rather weak animal.

Dogs, on the other hand, have been domesticated long enough to have lost their natural fear of people and can be dangerous to human life when they have too much freedom. Normally, both species try to evade direct confrontation with stranger-people. Responsible ownership is socially, morally and legally encouraged by the word *ownership*—it implies direct responsibility.

We can be unintentionally and unfairly judgmental about pets, often because we fail to see things as the other animal involved sees them. Pets tend to be very specific in rules, and this often fools their human companions into believing pets understand a great deal more than they actually do. To a pet, "no jumping on the table," as example, means, "no jumping on *this* table"—not the neighbor's table (or whatever). This is reflected most commonly in the astonishment of pet owners everywhere: "I can't believe he did that! He never does that at home!"

On the brighter side of sharing your life with a cat or a dog (or both), they make enchanting companions. Books on such companionships have been written on the joys of sharing life with a totally different kind of being (or beings). I've lived with at least one cat and one dog in conjunction with each other for so long, I can no longer imagine trying to choose one species over another.

Over my lifetime, I've had varied pairs of cats and dogs. And without a shred of doubt, I can say that each pair has made life more enchanting than the last pair, though individuals from any given pair have often overlapped. While there are always details about cats that apply (generally) to all cats, and the same sort of details about dogs that apply (generally) to all dogs, both species are made up of rugged individuals. (Too many people miss out on this facet of their personalities by approaching their pets as "just a cat" or "just a dog.")

Cats and dogs bring a sense of play and cheeriness in unexpected ways into the humdrum of daily life. There have been many times when my cat or dog has invited me to play, and I've been too busy. And after awhile, they'll play together—then I find I'm too busy watching them play together to get anything else done. The differences between cats and dogs make them interesting companions, for each other as well as for people. As long as neither one is allowed the upper paw over the other, they really like each other.

Cats and dogs are excellent teachers of nonverbal communication, as many people have learned personally. When anyone has a cat or a dog (or both) in the house, the indoor spirit is livelier. Often, such homes take on a gay madness, with toys flung in odd places.

Such homes usually have a lived-in and happy look about them.

Each kind of pet has uniquenesses. Cats play air back and forth across their vocal chords when purring. This activity is quite similar to people humming. People hum when they are happy; cats purr when they are happy. People sometimes hum when they are nervous; cats sometimes purr when they are nervous. It doesn't take a scientist to see that a cat purring at the veterinarian's is exactly the equivalent of a person humming at the dentist's.

Dogs are as unique as cats. Nearly every dog will bark and pass along the message to you (inside the house) that someone is outside the door. At the same time, they are passing along quite a different message to the person (outside the house) that there is a watchful dog alert to the situation. In rural communities, pet dogs pass along the fire siren, on those occasions when the sound is too faint for the people to hear. They perform this service to the community by howling in response to the fire alarm. I've talked to a rural fire volunteer who said he no longer bothers to call the fire station first when his dog starts to howl. He told me, "The dog's never been wrong yet!"

Living with other animals puts me in touch with the natural world in a way that would be beyond my ken without their presence. There are sights, sounds, and scents that interest cats and dogs that I would normally fail to notice, unless they eagerly point them out. They are aware of an entirely separate world mingled completely right within my own daily world, as they remind me constantly. My awareness of this gives me a sense of expansion, down to the smallest secret place inside me. And I've never found anything that can replace the affection they give me so freely.

Life with them has so many facets, so many unique experiences that cannot be achieved without them. Sometimes, we miss some unique experiences with pets, simply because they don't occur to us. For example, while many people think to go for walks with a dog, few people think to try it with a cat. It's quite unlike walking with a dog. It's somewhat tedious, but it's also enlightening. Cats pause at the edge of every open place, then hustle across it to the next bit of cover, where they once again proceed at a pace that allows an inch-by-inch examination of the terrain. It's so time-consuming, for me it remains an infrequent activity. But nothing else makes it so abundantly clear that a cat is a very small predator in a world filled with much larger predators, and careful travel is the cat's most important concern.

Going for a walk with a dog is equally rewarding. Anyone who has ever walked with a dog will understand what I mean when I say a dog follows you from the front. A freely running dog may never seem to look directly at the person he's with, but never misses a change in direction. It's interesting to note that dogs only truly follow from behind when they are unsure they are welcome on a walk. When a dog is certain he's welcome to walk with you, he'll take up his natural follow from the front position automatically.

Neither cats nor dogs actually enjoy following a person from behind in any traditional way. Dogs are perfectly willing to let the human pick the direction, generally; but cats definitely prefer picking the direction themselves. Cats will try to follow people if they think it's necessary, but they have no real ability to keep up with a long-legged human. (To *Felis domesticus*, a midget is a long-legged human.) When cats try to follow people and inevitably fall behind, there is a great deal of shouting, the feline equivalent of "Wait for me!"

As parlor psychologists suggest, there are unquestionably people who actually did originally acquire a pet as a substitute for something else. But since they've acquired one, most of them have learned that the satisfactions of pet ownership are unique. In no way are pets a satisfactory substitute for anything else.

Meantime, there are at least millions of people who keep pets because they want pets—not substitutes. People are only just beginning to understand how functional pets really are within the human community. As a result of their study of human/pet relations, Doctors Beck and Katcher wrote *Between Pets & People*, in which they listed the following as the *minimum* benefits of pet ownership:

1. They provide companionship.
2. They give us something to care for.
3. They provide pleasurable activity.
4. They are a source of constancy in our changing lives.
5. They make us feel safe.
6. They return us to play and laughter.
7. They are a stimulus to exercise.
8. They comfort with touch.
9. They are pleasurable to watch.

Pets affect people's lives in many ways, and the ways are often subtle. I find it impossible to become too introspective when my pets are around. No matter how emotionally loaded a situation becomes, I find I can step outside my own perspective and see the situation a bit more objectively simply because of their constancy in my life. I know that my privatemost life, shared with my pets, simply isn't going to change because of any given situation that is external to that relationship. My pets are always thoroughly convinced of my invincibility, my morality, and my survival. And, like any other positive association, their convictions rub off onto me and renew my faith in myself when it's flagging. At times, this has been no small achievement on their part. They are so sure of my capabilities, they must be right. (And how could I let the little guys down?)

In return for a small amount of care, pets give people a good deal of emotional support, often in intangible ways. And much of the time, people's basic attitudes about everything are subtly altered by caring for pets. There's a subconscious adage among pet owners everywhere: "Anyone who voluntarily takes the time and thought to care for another being can't be all bad." And marvelously, this seems to be true. Anyone who can extend their concern outside of themselves, outside of their relatives, all the way out as far as a being of an alien species really can't be all bad, by anyone's definition of bad. Because pet owners tend to approach other pet owners with this basic assumption as a starting point, pets often create friendships by serving as a bridge across class differences, across religious boundaries, between racial opposites—separations it would normally be difficult to span without the intermediary of a common interest—in this case, a like pet.

There are few starting places for compatible conversation between otherwise separate people as there can be between two (or more) cat lovers or two (or more) dog lovers discussing their pets. When simply owning a pet makes it possible to cross staid boundaries, pet ownership has the capacity to open conversation, to understand many things from a wholly different point of view, supplying insight into others' lives in ways that might otherwise be unattainable.

In recent years, I've been making an effort to closely observe the human's reaction to meeting my pet. Most of the time, pet owners focus wholly on our pet's reaction to the stranger, not the other way around.

I've found this way of observing such exchanges an enlightening experience.

In the larger society in recent years, it has become less common for people to ask permission to touch your pet before doing so. This new attitude is both unwise and, occasionally, downright dangerous. The very commonness of cats and dogs gives too many people the idea that all of them are safe to handle, anytime, anywhere. (For those who don't know better, pet owners should have the absolute right to tell you *first* if it's really all right for you to extend your naked hand out to Fluffy or Fido.)

Be that as it may, the interesting part comes when the stranger actually touches your pet. Many cats have an innate dislike to being touched by people they don't know. Less commonly understood is that many dogs feel the same way. Ninety percent of the time, the initial touch (usually a pat on the head) serves most clearly as a dismissal of the cat or dog. Observing carefully, I've noticed a faint aura of relief from the people who wish to touch my dog. Once she's been touched, she is usually thereafter ignored. I've come to interpret this as the *petter* deciding on a subconscious level that he (or she) is safe, having satisfied themselves that my cat or dog intends no harm.

Perhaps 10 percent of the time, people who wanted to pet my cat or my dog are seriously interested in the being. Neither of my current two really like being handled by strangers, but it's interesting to notice how frequently the person involved doesn't really care about the other animal's opinion.

Some people of course do care, greatly, and respect the animal's wishes. Those people who don't care (or notice) seem to expect the cat or the dog to automatically suffer the undesirable attention. In any such confrontation, there are at least three minds present: the person who wants to do the petting, the cat or the dog, and the owner. And by this time, I've observed enough other pet owners to know I'm not unusual in siding with my cat or dog in any minor effort to avoid being touched by a stranger. Pets should have an absolute right to accept or reject petting and they do decide for themselves.

The majority of pet owners will take the pet's side in such confrontations, without hesitation. On a one-to-one basis, no one can know the real you better than your pet. Pets study their people with an intensity that no other person can possibly achieve. Pets understand their own people on a level that even pet owners find uncanny. There is simply no other being who cares about individuals as much as their pets do. They bet their very lives on you, every minute of every day.

I have several friends who believe that pets can read your mind. After long debate, I've come to believe these good people are largely right. My every movement, subconscious action, unconscious gesture are levels of my being that I often fail to notice. But those details aren't missed by my pets. Perhaps I always scratch the left side of my head, get a drink of water, or otherwise demonstrate a certain level of restlessness just before I suddenly (to me) decide to go for a walk. But my gestures aren't missed by my pets; they do notice them. In that precise sense, my pets do read my mind, usually far ahead of my noticing what is passing there.

Understanding that, I'm less surprised that my cat always seems to know before I do when I plan to give up writing (a dull, dull activity that demands naps) for an occasional half hour. She seems to know this long before it occurs to me that I've written myself into a corner, that I need a rest from whatever I'm working on at the time. I can be expected to take a book and my cat out onto a bench in the yard, where she can expect to be petted and tickled while I idle away a brief time.

The moment the thought of a walk occurs to me, seemingly a brand-new thought, I glance at my dog and she knows my plans. At times, it's downright eerie, but most of the time it makes me feel secure and happy to know I'm so understood.

No human can get as close to the emotional me as my pets. It bears repeating that no other person is willing (or capable) of spending that much time to study me that thoroughly. Pets are often artfully in tune with their people, often using their senses in ways that are beyond people's ability to interpret. Pets come to recognize a certain tone in their people, a tone that indicates clearly to them that their person would rather be doing something else. Konrad Lorenz spoke of his frustration with a bitch he owned, who always knew ". . . which people got on my nerves and when. Nothing could prevent her from biting, gently but surely, all such people on their posteriors." Dr. Lorenz got to the point where the instant he felt annoyed with anyone, he would reach for the bitch, only to find himself too late—she'd already bitten the offender.

Your pets study you. My pets study me. My pets know every detail of my daily and weekly and monthly schedules. They know even better than I do when I'm settling into a chair if I plan to stay or pop up again. They almost never misinterpret my moods, my intentions, and my interests. Pets are *aware*.

Yet in spite of pets' genuine awareness', I've never been convinced that pets understand people. *My* pets understand *me*. *Your* pets understand *you*. But switching any of the members of these exclusive groupings around doesn't make for understanding. In an odd way, this supplies people with a perspective on pets that pets cannot hope to equal. To some degree, motives behind human activities always remain an insoluble enigma to pets. While my pets have absolute faith in me, they don't have much faith (if any) in anyone else. And the quality of their faith in me fairly demands equal faith in return. And there lies a difficulty.

Equal faith is not always easy to give pets. Sometimes, it's not even possible. From long association with people, pets learn to lie. Over time, pets come to understand that items they feel should be valueless are, for some unfathomable reason, valued by their people. The puppy that has chewed a leg off the sofa while you were out gives you the warmest possible greeting upon your return. From that puppy's perspective, how could it possibly matter to you that a mere sofa was damaged, when you get such a genuine, effervescent greeting when you return? Continuing with the puppy's view, the globe began spinning again when you arrived home; what could possibly be so serious as to change the significance of your arrival? And then...the incredible truth comes out: a sofa is important to you. You actually perceive it as *important*.

Eventually, all pets learn that, astonishing as it must seem to them, you do value odd, inanimate things like sofas. The point that's hard for people to remember at such times is that the greeting of incredible warmth is no less sincere, in spite of seeming to be staving off the predictable punishment. Certainly, the effervescent greeting seems vastly overblown because of the poorly hidden lie it contains. But the puppy really is glad to see you. The fact that the enthusiastic greeting only temporarily defers the expected punishment really is a side issue.

Like dogs, many cats develop a basic understanding of how to lie. A cat may be sitting in a flurry of yellow feathers and seem to express total astonishment that there was a canary in the house when you left and there doesn't seem to be a canary in the house now. ("Really? A canary? What do they look like?") Both cats and dogs proceed predictably, from the incredibly transparent lie through the astonishment stage, onward to the "I am only a small helpless animal" stage. But there is a far bigger issue transpiring in these exchanges, inherently complimentary to people.

When cats and dogs try to lie to their people, it's the ultimate flattery. Pets often know perfectly well what their people do and do not want them to do—but those rules bear little relation to their own value system. Yet they will try to lie *just to please their people*. Lying is a wholly alien concept to most animals, yet pets actually try it. The fact that such lies are usually transparent (to put it mildly) doesn't diminish the demonstration of intelligence that starts the whole thing. While temptation may lead any pet astray from time to time, pets work quite hard at maintaining their people's high opinion of them. That some pets will even emote guilt and sadness over some indiscretion that is unfathomable to them is ultimately flattering, and speaks volumes for the quality of the relationships between the pets and their people.[21]

There's another side to this lying issue, of course. If a pet's lie is transparent to his people, the reverse is also true. *People cannot successfully lie to their pets.* Trying to lie to a pet is the ultimate waste of time. It would take the best poker player in the world to convince a cat or a dog that lives with you, that you are not going out—especially when you are. Both species are fully aware of your attempts to lie. If you share a favored activity regularly with your cat, don't expect to be able to lie successfully to your cat and do that activity without him; he'll know perfectly well that you're lying to him. There isn't a dog in the world that doesn't know his owner went out for a walk—*without his dog*. When pets lie, their people usually know it. And the reverse is just as true: pets are fully aware of any attempts their people make to lie. An owner's efforts at hiding the truth from their pets are as wasted as the reverse.

Pets and their people are an important reflection on humankind. Individual relationships represent only a fraction of collective human involvement with other animals, but that fraction is significant, as it helps to comprehend, on a deeper level, some of the effects that animals have on humanity. Humankind's personal relations with individual pets provide people with insights into the development of intelligence. Mutual concern seems to develop a special kind of intelligence, a different sort of interspecies communication, a unique sense of heightened awareness to which people might not have any other access.

On the less substantial, more slender level of a person's individual relationship with her (or his) pet, such relations are proving more significant than was ever suspected before. What people receive from their pets is subtle, but nevertheless substantial. Whether a person enjoys low or high social status, low or high income, their pets react the same. Pets don't care what religion their people might be; they don't even care if their people practice their religion. Pets know all the faults and weaknesses of their owners. Pets know much that matters— really matters—and they don't care about anyone else's opinion.

My pets know whether I wake up surly or sweet. *Your* pets know when you don't mean what you say. Even when pets only get the general idea of our words, they are exquisitely aware of the emotions behind those words. Pets will share willingly lows and highs and they do this sharing in hovel or castle, in famine and feast. Pets accept their people's frivolous moments without expecting them to last long, as well as gloomier moments—pets share it all and are simply glad of the sharing—bad or good. Collectively and individually, pets may very well be the best friends we'll ever have.

According to Doctors Beck and Katcher, thresholds are very significant between pets and their people. It is the place where people value their pets most highly. The essence of that threshold meeting is that you are home, you are in a safe place, and you are quite welcome. No matter how the world has treated you today, you had the power to brighten the whole world for another being just by going home. You will be loved and valued here.

CHAPTER 12
PEOPLE & OTHER ANIMALS

One of the subtle differences between people and other animals is that people assume sovereign authority over the whole of the animal kingdom. No other animal does that. Since we do assume that authority, it's less surprising that we try to exercise it from time to time. Whether or not we really had that authority has constantly been reflected in our lack of success.

A surprising example of this is the North American bison (*Bison bison*). According to the Audubon Society: "The destruction of the Bison began about 1830 when government policy advocated their extermination to subdue hostile tribes through starvation . . ." In other words, the U.S. government was aiming at the Native American peoples indirectly by encouraging the decimation (slaughter) of the great bison herds.

In seventy years' time, the remaining bison could be counted individually. And when they were counted, there were less than a thousand. (Incidentally, the bison is the best example of a near success of the deliberate eradication of a species.)

While the decimation of the bison unquestionably did succeed in subduing Native Americans, it had a hidden expense attached. As the bison were killed in large numbers, the prairie dog population grew. An intense campaign to reduce the prairie dog numbers also had an unexpected expense: the near-extermination of the black-footed ferret. There is every reason to believe that other imbalances (and probably extinctions) occurred during that time, and simply went unnoticed. All three of the obvious species were saved, but it was quite close. Of the three, the black-footed ferret is still in grave danger.

These facts present an example of the unexpected complexities of the fabric of life itself. Meddling with individual strands of that fabric has taught people the hard way to be more cautious than was believed necessary in the past.

I was only four when I met Bobo. I never knew how old he was, and I have no idea if he is still among the living. But I never gave up the interest he encouraged in me on that long-ago day: what is the difference between people and the rest of the animal kingdom? By now, I feel I know.

At that time, many years ago, I set my mind to learning as much about animals as I could. By the time I researched this book, I already knew a great deal about other species, yet I could not, not with any certainty, answer that basic question. I knew lots of isolated facts, like that a giraffe's heart weighs twenty-five pounds; I knew that the word *mouse* comes from Sanskrit and means *thief*; I knew that a race horse's resting heart rate was about twenty-four beats a minute but, in a race, the same horse had a heart rate of about two hundred and fifty beats a minute. I knew a lot of little isolated facts like that.

I understood a great deal more: I understood that many animals have moveable ears as a way to listen all around; I understood that migration is where an animal population shifts for a season, and that emigration is where animals shift from one location to another for life; I understood that unknown animals are still being found on a regular basis; I understood that rabies is a disease that can be as horribly fatal to me as it might be to my cat or dog or any other mammal.

Still, I didn't know or understand much from the perspective of the things I know and understand now. While I had a tacit understanding of how important the cow is in our everyday lives, I didn't understand the enormity of her participation in matters of a global scale; I understood that many other animals, besides people, are considered intelligent, but I had no real understanding of how little we know on the subject of intelligence; I had no real idea of how badly humans have failed in trying to communicate with other animals in any way past the mundane; I was perpetually confused about how varied other animals' eyesight may be; I had heard all my life that human population was escalating at a dangerous rate, but I hadn't a clue as to how much trouble we really are in. Now I understand that this population problem is universal, involving literally billions of people.

I have also heard, throughout my life, about how rotten the human animal can be, how reliably horrible any of us might be. Needless to say, these words always came from a human mouth. When I first heard it, I was young enough to wonder if the person speaking was really telling me about himself. Since then, I have come to understand that such statements mean everyone *else*, but never the speaker. I have also come to understand that I am a part of that everyone, and I have learned to accept that judgment. I've even learned to enjoy that judgment. I am a part of everyone. (It has a nice ring to it.)

And what I finally came to understand is that all I had known before were little isolated facts, that I had understood small isolated pieces of a subject that encompasses just about everything, as far as what John D. MacDonald once referred to as this "whirling mud ball." I had no deep understanding of the universal realities of life, the universal version of life that involves time in ways that I can barely imagine. I had no idea, not really, of how much a part of the fabric of life itself that all those little isolated facts were.

Now I do.

And I have come to understand that we are in big trouble from the usual source of our biggest trouble: ourselves. It took a great accumulation of little isolated facts before anyone began to see the larger picture of life; that life itself may actually be of a single integrality; that life itself is made up of trillions of integral parts, each smaller than the big picture, but obscurely significant to the whole: the biosphere. This may be the greatest discovery of humankind yet. All the evidence isn't in, but evidence has been accumulating, mounting in both solidity and volume.

To me, the most noted developer of this hypothesis is J.E. Lovelock, a Fellow of the Royal Society of Great Britain. He entitled this way of looking at things *Gaia*, based on the Greek word for what we call *Mother Earth*. In his published work by the same name, he offered the postulation ". . . that the physical and chemical condition of the surface of the Earth, of the atmosphere, and of the oceans has been and is actively made fit and comfortable by the presence of life itself." It was only yesterday that the much smaller focus of life was the accepted explanation, which held that ". . . life adapted to the planetary conditions and they evolved their separate ways." Lovelock's hypothesis is a subtle but complete reversal of that former (and much smaller) view of life.

The Gaia theory isn't an explanation of life. It's much simpler than that. It's what works. Grossly encapsulated, Lovelock's view is that the biosphere itself is a life and that *all* life within that thin layer (of the biosphere) is inter-supportive; that the life on the planet is the reason behind the stability of the planet: atmospherically, geologically, universally.

If Gaia turns out to be an accurate analysis of life, then all human divisions, real and imagined, may not matter a whit in the long run. It's far more important that life itself needs air, water, food, and living space, than whether the life needing these things is human or not. As human options diminish, often due to past errors, it becomes clearer that life itself needs the genetic diversity of the mosquito, the unrecognized services of the coyote, the adaptiveness of the rat—just for the biosphere to stay alive and healthy.

In the recent past, humanity assumed authority over all other life, and that assumed authority was freely exercised. But instead of broadening human options, it narrowed them dramatically. One result of that attitude was learning that there is indeed such a thing as sovereign authority—and we people have it.

To date, no other animal, individually or collectively, has shown any sign of challenging our sovereign authority. So far, there appears to be no other beings on the globe who desire the position of power that people hold. There may never be such a being. There is no reason to believe, however, that our authority is such a desirable position, no reason to believe that it's yet another reason to pat ourselves on the back.

Oh yes. Our sovereign authority is real, no mistake there. It's real largely because we assume it is. Real authority and assumed authority are, in essence, the same. So simply exercising authority makes it an acquired reality.

In this case, it's painfully like holding the proverbial tiger by the tail: any number of unpredictable events—a monumental earthquake, a collision with a large meteor, or simply a newly evolved bacterium—any sort of natural development may turn out to be a monster that people have no real control over. Any such unanticipated event may forever alter the situation, resulting in mortal wounds to the biosphere itself. The ugly reality is that human control, while real, is gross: that is, we can indeed destroy all other life or preserve all other life. And it looks like the preservation of as many other life forms as possible is the only option left open to us; we no longer have the option of random destruction, if we ever really had it. Those days are past.

There are unquestionably people everywhere who have yet to

comprehend this reality. Such people live by the happy-go-lucky fable that nothing has changed—not really. Such folks are inclined to state that the world is the same; you'll never change it. To such people, I must say: Well, no! The crust of this planet has never had this many billions of people on it before. And while I, personally, may never change the world, it has already been changed, irrevocably. And the entire line of thought is, after all, just another variation on the theme of precedence: "We have always killed other animals off. Why stop now?"

Historically speaking, people are a well-established species that has suffered few setbacks. Of those setbacks, some have been colossal at times: famine, drought, and plagues. People will unquestionably suffer setbacks again in the future. The bottom line is that people are still here and a functioning part of the biosphere.

Speaking geologically, however, humankind is the new kid on the block. As such, humanity is forced to operate in a game where the rules are best understood only after errors are made. People are intelligent animals, certainly intelligent enough to be aware that others have ruled Earth in the past, such as large dinosaurs, just as people rule now. But those others were here and they are not here now. And all we can do is guess why those other rulers disappeared from the ongoing game of life. Why are they gone? Did they make some fatal error that we haven't discovered yet?

Dinosaurs were once the ruling class on this planet, and most of them are gone now. Why they aren't here now has been a long-asked question.[22] No matter which of the many hypotheses is the actual truth, the largest dinosaurs were knocked out of the game of life forever about 65 million years ago.

People rule Earth now. Often, that rule is more of a blind groping, although it has nearly always proceeded with arrogance. But people are learning. The lessons we've learned aren't always pleasant, and there have been times when the answers were obscure. But we're trying—and succeeding quite often.

We've learned that it's the rate of extinctions that makes for so many of our current environmental problems. Some time ago, J.J. McCoy wrote, ". . . one-half of the known extinction of wildlife over the past 2,000 years occurred in the last fifty years." Understanding that intellectually is separate from the emotional shock of watching a film of massive herd migrations and suddenly realizing some of those herds no longer exist—except on film.

When this thought first struck me, it brought a sharp and painful awareness of what it must have been like for those who had watched the great flocks of passenger pigeons, then realized one day that those pigeons no longer existed. I felt as observers must have felt when they witnessed the senseless slaughter of the seemingly inexhaustible bison herds, the profound astonishment of seeing something so impressive come to an end that was hardly credible.

Human sensibilities have been reached, and none too soon. Our sovereign authority has become a roller coaster ride of unexpected successes and (sometimes) dismal failures. When I again tried stepping back for a more objective view, I stumbled onto something unexpected.

More than forty years ago, when people were beginning to understand the importance of our mutual global environment, the whooping crane was the center of their attention. And it's that same species, the whooping crane, which best describes our current environmental situation.

The elegant whooping crane was never very numerous. The bird stood more than five feet tall, and mature birds had pure white bodies. When the cranes flew, their wings spread more than seven feet, with jet-black wing tips. Whooping cranes became an irresistible target to many people; cranes also suffered continuous habitat disturbance on the ground. By the late 1800s, the birds were already in serious decline. By 1941, there were only fifteen cranes left on the North American continent.

Their plight was desperate. In reaction to it, the U.S. Fish & Wildlife Service, the Canadian Wildlife Service, and the Audubon Society joined forces to create the Cooperative Whooping Crane

Project. Yet in spite of their best joint efforts over the next nine years, the number of cranes only rose to twenty-one.

By 1965, global interest in endangered species had grown by quantum leaps, all around the world. The U.S. Congress voted for $350,000 to begin the Endangered Wildlife Research Program. It was the whooping crane that was the first species on its list.

Was all the focus on the whooping crane successful? Well, the whooping crane still stands over five feet tall; the whooping crane still flies with a wingspan of over seven feet. And the cranes do this in numbers greater than were ever expected. Today, there are more than 150 whooping cranes.

What I realized was that I'd assumed the whooping crane had become extinct. The cranes were the focus of a big fuss when their numbers were tiny. Their low numbers were newsworthy; their higher numbers were achieved slowly and steadily. Slow and steady increase is not news. So finding out about their recuperation made me realize on a deeply personal level that any extinction is not a foregone conclusion.

Many species that have spent time on the endangered list have been preserved through human effort. As only one example, the northern elephant seal spent some time on that list. Many males reach twenty feet in length and can weigh nearly four tons. They still reside off Baja California's west coast. But there was a time when there were less than one hundred of them. Mexico granted them protection in 1922. And directly because of Mexico's forethought, there are more than 100,000 of those seals today—and all of them are descended from that original small group. The point is, their extinction was not a foregone conclusion.

The primary cause of extinctions was stated most succinctly by naturalist Willy Ley in 1941: ". . . a species that is too strictly adapted to a given set of conditions is bound to face extinction if, as and when these conditions change." From that perspective, I can understand why there have been so many recent extinctions. And I can also see why the *rate* of those extinctions keeps speeding up.

It's a bad enough problem, all by itself. But there is a bigger problem that is adding to it. According to biologist Paul Ehrlich, "No longer are more species being created than are going extinct each year..." Unfortunately, this is yet another first in the long history of the planet. There are too few new life-forms coming up to replace those life-forms that are rapidly going extinct now. Therefore, the overall reduction of the variety of life we share this planet with is likely to be permanent.

Understanding of this vital point must grow rapidly. People are thinking in wholly new, innovative ways about how to solve problems within the natural order of the environment. No one expected human concern about the environment to grow as rapidly as it has; no one expected that awareness to quickly change the thinking of the average person—and it has.

In the summer of 1985, a clever person at a Montreal radio station played the sound of a female mosquito over the airways, succeeding in attracting droves of male mosquitoes to radios all around the city. The males, who don't bite, were dispatched without toxic sprays.

When the French military had too much interference from great flocks of invading sea gulls on their runways, they organized the French Falcon Corps. The Corps maintains ten falcons who successfully keep away an estimated fifteen thousand gulls.[23]

This concern of people about the environment isn't exactly brand-new. According to Paul and Anne Ehrlich, the Thames was still a good fishing river in the early 1800s, but by 1850, all commercial fishing there had ceased because of water pollution. However, today that river has large numbers of fish—and even waterfowl. That only happened because people tried to reverse the trend and succeeded!

Great numbers of species have been expected to make the trip from the *endangered* list to the *extinct* list. But a surprising number of them have instead made the reverse journey—off the endangered list entirely. While the endangered list is still long and growing in some places, and the extinct list continues to monotonously grow, the number of species being removed from the endangered list is also growing. Perhaps it isn't growing as fast as we'd like, but it is at least growing!

Within a short time, the Arabian oryx was removed from the endangered list because of the efforts of public and private zoos. A small herd was even returned to their country of origin.

The American alligator took a bit longer, but is now off the endangered list, as is the California bighorn sheep—currently being re-spread in his former range. The beaver is today more numerous on the North American continent than he was at the height of early fur trapping.

There have always been some animals that had high enough numbers for them to actually adjust to new environmental pressures. The Pepper and Salt moth (*Bistion betularia*) of Great Britain was at one time commonly light in overall color, to blend properly with the lightness of the tree trunks—their favorite resting place. Among the light-colored moths, there were invariably a few darker moths, which were easily visible to hungry birds.

In only fifty years, this moth changed from generally light coloration to generally dark coloration in reaction to air pollution. As the tree trunks darkened as a side product of industrialization, the survival conditions for this species of moths exactly reversed themselves in that brief time. Known as *industrial melanism*, this color change has occurred in other species as well, for much the same reason. But the Pepper and Salt moth went from light to dark coloration within a single human lifetime, making it the most notable change.

Not all species have high enough remaining numbers to make such adjustments. There is a small marine moss animal (*Bugula neritina*)[24], growing on the moorings of piers in southern California. Only discovered in 1981 by Dr. Thomas Eisner, he promptly placed the animal on the endangered species list, as these animals are not adapting well to recent water pollution.

A global sense of community has been growing around the human interest in other beings. People who are interested in animals, wild or domestic, are more alike than different, no matter what their political or tribal divisions. People have learned that there are no political divisions between the other animals. The result has been a developing sense of community among humanity. As the issue is global, so is the solution. And that is how people now approach it.

In this same way, the kingdom of animals can no longer be neatly or easily divided between wild and domestic. Far too many people have now seen wild capybaras grazing alongside cattle in South America or wild elephants chasing cattle off their reserves in India to miss the implication. Living space has become too crowded, too scarce, too populous for any such simple divisions between wild and domestic to be realistic any more.

It has become ridiculous to only speak of damage done to the biosphere in the past, as it has become ridiculous to only speak of solutions in the future. There is very much a *now* in the middle there. And while it's common to hear we must take good care of Earth for our children, *now* is as vital an issue as is the future. We tend to forget that we have already inherited a lesser Earth than our parents did, and a far lesser Earth than our grandparents knew. *Now* might make the important difference as to whether there will even be a future.

Today, millions of people belong to environmental groups. This global concern is most remarkable in that it transcends all the commonly perceived barriers of ideology and geography.

All the sciences have been reporting to humanity for a very long time now that we're living on the rim of biological disaster, that human existence itself may ultimately depend on human actions today, that the foundations of human superiority have been built upon the thinnest of ice and we've been salting that ice for traction.

With that reality in mind, I'd like to encourage scientists everywhere to stop focusing only on the negative side. Beyond a doubt, there is bad news right and left on the global environment. However, people have learned, often in painful ways, that the overview of the environment is, at best, shaky. People are aware that a single past attempt to wipe out the mosquito (with DDT) damn near cost us the entire category of birds—very much including the proverbial bluebird of happiness.

Humankind has a stupendous capacity for learning. I don't believe, even for a minute, that any scientist practicing his craft forty years ago actually expected people's attitudes toward other life to change so rapidly.

As an authority, the scientific community tends to forget that when people held the idea that humans were separate from nature, scientists also held that same view. And as the scientific view of the world changed, so did the view of most people. Science has small faith in people. That small faith has not always been justified. The relatively recent change in attitude toward all other life has been global.

The phenomenon has been worldwide. There is now a multi-million dollar industry in the state of California where people pay to be taken out in boats, only on the chance they might see a whale. There are more than seven hundred wildlife clubs in Kenya today. And in India, where living space has long been at a premium, many families have given up their ancestral homes to provide land for something they felt more important: tiger reserves.

I think there's ample evidence that people everywhere exhibit genuine concern for other animals.

Marlin Perkins (1905–1986) of Carthage, Missouri, was for many years the host of the television program, *Wild Kingdom*. His closing remark to each episode of the program was always the same: "*Man is the ultimate ruler of the wild kingdom.*"

For a lot of years, I wanted to meet Mr. Perkins to argue over that remark. It always seemed to be such an anthropocentric (man as center of the universe) view of the world. His oft-repeated comment seemed to disregard the sheer biology of the situation. And it certainly didn't allow or encourage human contemplation of the millions of animals on Earth that don't even suspect the existence of all-powerful humankind.

Upon hearing the news of Mr. Perkins' death, I found myself often thinking of his confident premise. For the first time, I could see a way he might have been right.

Every overlord—king, prime minister or president—is essentially ruled by followers. No leader remains in power long, without surrendering to the wishes of the subjects beneath him (or her). The whole process demands cooperation of a sort, on both sides. No leader is a leader without followers; no follower has any focus without a leader.

And with this new sort of trial humanity is facing to preserve the plants, foster other animate life, encourage the function of the biosphere, it is *absolutely essential* to put all our faith into the hands of the poor and the rich, the ignorant and the educated, the child and the adult, the immature and the mature, the religionist and the atheist, the blacks, the browns, the reds, the yellows, the whites among people. It is *absolute reality* that humanity's future rests in the control of the varied political nations of Earth, whether they are war-torn or peaceful, whether we like them or not. And we'd better stop thinking of some of them as third world nations, and face up to the reality that there is only *one* world and we all share it. In short, continued human existence rests in the hands of the "ultimate ruler of the wild kingdom," the common people.

Potential threats exist everywhere, from the smallest child with a BB gun to the largest adult with a rifle, or an adult with a finger poised over the proverbial button or simply a dump truck full of toxic waste. The fate of humankind, as a species, rests in every human hand. There can be no more potshots at the eagles, no more beer cans casually tossed into lakes, no more industries dumping waste.

It will take the cooperation of nearly everyone to reverse the current trend of extinctions. This isn't just a matter of losing a few species: the rate of extinctions threatens us, one and all. Human survival, as a species, is dependent on this planet's environment, in ways that we are just beginning to understand.

After years of studying this issue, I still don't know who is closer to our everyday reality: Those who believe *no comparison possible* or others, with their *no essential difference* approach. As I see it, it is realistic and reasonable to make comparisons, and there are indeed some essential differences between humankind and the rest of the animal kingdom. But I've also come to understand that either extreme position doesn't really matter; I've come to understand that we are far more like other animals than different when generalizing, and that those similarities far outweigh any differences.

And I've also come to understand that there is no *us* and *them* between people and other animals. The modern world is filled with people who understand that humankind stands proudly *among* the animals—neither above, nor below, nor separate—but quite comfortably among them. And, while the qualities of our companions are varied, they are always high. The rest of our kingdom delights and enchants us at every turn: they provide reflections of ourselves that we cannot see in any other way. And we're working hard everywhere to make certain we can go on sharing this world with such other beings forever.

We've lost some. We'll lose more. But we're also succeeding quite often and ultimately, we keep trying.

We can't lose.

We really can't.

ENDNOTES

1. There is substantial evidence that this form of vaccination was long utilized in Africa and Asia prior to Dr. Jenner's publication.

2. The chimpanzee's use of tools had an odd confusion for me. I had always credited Jane Goodall's work in the Gombe Reserve in Tanzania with this discovery, though she had occasionally mentioned that there had been such information earlier. And, in my digging around on the topic, I found the use of tools mentioned quite casually in Willy Ley's work of twenty years earlier.

3. Rods and cones in the retina are receptors for the optic nerve.

4. In the *dive reflex*, the heart rate, as well as other metabolic processes, slow down drastically.

5. The opposite cross, a male horse [stallion] and a female donkey [jenny] is an entirely different creation—a hinny. The hinny looks a great deal more like a horse than the mule does.

6. Walkabout: an Australian aboriginal walking tour, usually done alone.

7. In the world-famous story, *Bambi*, written by Felix Salten, Bambi was trying to *protect* Faline from bucks who were competing with him for her attention—quite a different, though more realistic, view of events.

8. In a single week, on TV, I heard it stated that India has the fastest growing population on the globe; later, I heard it was Africa; late that same week, I heard it was China. What possible difference could it make, overall, as to which nation is doing the most damage since we are *all* contributing?

9. According to Vitus B. Dröscher, one hundred million is not an arbi-

253

trary number; it's the point at which sexual reproduction causes the tempo of evolution to speed up. Again, the number is relative to other factors.

10. The dog is always exempt from such estimates. Some authorities estimate the dog's domestication as far back as thirty thousand years. The estimate is always hedged because, while canine bones are often found in archeological digs, the question becomes, "Was the dog a resident? Or was the dog a meal?"

11. Wild Boars are classified as *adventive fauna* in North America; they neither migrated nor evolved here. This automatically means they are destructive to native flora and fauna, to some extent. Wild boars are even more destructive than most introduced species, killing many native snakes, birds, rabbits, deer fawns, others. They are devastating the native peccary.

12. Interracial and interfaith marriages are still generally uncommon around the world.

13. The name *woodchuck* is derived from a mispronunciation of the Cree *wuchak*, a name that covers this (and several other) animals. The name has absolutely nothing to do with their behavior.

14. What Farley Mowat discovered was a series of changes decimating the deer herds. I suggest reading *Never Cry Wolf* by Farley Mowat, copyright 1963.

15. Few of the plants cultivated today can be correlated with existing wild plants with any certainty. One of those few is the rangy *Queen Anne's Lace*, a scraggly plant with a thin, off-white root. Eventually, through the efforts of thousands of people, this root has been developed into the huge orange carrot of today.

16. Here quoting John C. McLoughlin, "As one Fuegian put it, on being questioned about his people's eating their women but sparing their dogs during hard times: 'Dog catch otter. Woman no catch otter.'"

17. Aquinas' general theory was that those who exercised cruelty against lesser animals would acquire a taste for it, eventually becoming cruel to people.

18. Such reports lack the rate of speed the locusts flew at, as well as whether they had a head wind or a tail wind, which could change their estimated numbers significantly.

19. Current theory is that birds do indeed have a smell sense.

20. Along the lines of the parlor psychologist's view, I'm certain I've met at least a few people who had children as substitutes for pets. Such logic is clearly a two-way street.

21. I must refer back to an earlier mention of the dubious gift people have given to dogs: guilt. As social animals, dogs already had a tendency to feel guilt, but we've honed it to a fine edge over the millennia. Entirely too often, I've met dog owners who believe their dog knows what he did wrong days ago, because he looks so guilty each time his owner glances his way. While humanity has sharpened the dog's natural tendency to feel guilt, we've given them no way to understand the why behind it. The guilty look is a direct extension of the dog's ability to perceive what his most important person feels about him. He doesn't have to be a genius to know he has fallen out of favor with the highest being in his world. And he probably hasn't a clue as to why his personal god is angry. In such circumstances, the best of all worlds is to punish swiftly, then drop it. Dogs know their people well; certainly well enough to know if those people tend to be vindictive. And they still love those people anyway.

22. Nobel Prize winner Luis W. Alvarez, his geologist son Walter, and several others suggested one of the most recent hypotheses of what caused the disappearance of the large dinosaurs. Their hypothesis is that Earth was struck by asteroids about sixty-five million years ago; that the collision caused an enormous dust cloud that blocked the sunlight for so long, the temperature of the planet itself dropped; one result was that much of the plant and animal life on the planet perished.

23. Flocks of gulls come in, and a falcon or two is released. Gulls don't hang around falcons; they find somewhere else to go and quickly.

24. Commonly mistaken for a sort of seaweed, these animals (*Bryozoa*) live in colonies that resemble mat-like debris. Microscopic as they are, it will surprise some people to learn that they are the focus of a good deal of human attention. This rather obscure animal has been found to contain an unexpectedly important substance called Bryostatin. Bryostatin happens to be a powerful anticancer element. And this particular bryozoa is one among many on the endangered species list.

BIBLIOGRAPHY

INTRODUCTION

Ley, Willy. *The Lungfish, The Dodo & The Unicorn.* Viking (1941), p. 310.
Sanderson, Ivan T. *Living Mammals of the World.* Doubleday (1972), p. 102.

CHAPTER 1

"ABC 20/20," Show #525 (6/20/85).
Audubon Society Field Guide to North American Birds, The. (Eastern Region), Alfred A. Knopf (1977), pp. 511–512.
Audubon Society Field Guide to North American Mammals, The. Knopf (1980), p. 18.
Book of Popular Science, The. Grolier (1924–1931), Vol. 1, pp. 223–229, Vol. 6, pp. 2020–2022, Vol. 12, pp. 4568–4570.
Curtis & Guthrie, eds. *General Zoology.* John Wiley & Sons (1927), pp. 209–210, pp. 238–240.
Collier's. Funk & Wagnall's Co. (1946-1956), Vol. 1, p. 57; Vol. 2, p. 1190.
Encyclopedia Americana. Grolier Inc. (1984 (& 1927 on)), Vol. 12, p. 704.
Hickman, Cleveland P. *Integrated Principles of Zoology.* C. V. Mosby (1966), p. 544.
Johanson, Donald, and Maitland Edey. *Lucy: The Beginnings of Humankind.* Warner (1981), p. 31, pp. 320–321.
Ley, Willy. *The Lungfish, The Dodo, & The Unicorn.* Viking Press (1948), pp. 1–14.
McClung, Robert M. *Lost Wild America.* Morrow (1969), pp. 108-111.
McNeill, William H. *Plagues & Peoples.* Anchor (1977), p. 222.

257

Nelson, Robinson, & Boolootian, eds. *Fundamental Concepts of Biology*. John Wiley & Sons, Inc. (1967), pp. 3–12, pp. 256–257, pp. 292–293.

Sanderson, Ivan T. *Living Mammals of the World*. Doubleday (1972), p. 242.

Scott, Jeff. *Bird Watchers Digest*, "Winter Spectacle." I-90, P. O. Box 110, Marietta, OH 45750.

Ward, Henshaw. *Charles Darwin & the Theory of Evolution*. New Home Library (1927-1943), p. 19, pp. 23–41, pp. 55–64.

Waldeman, M.D., Robert H. and Ronica M. Kluge, M.D., eds. *Textbook of Infectious Diseases*. Medical Examination Publishing Co., Inc. (1984), p. 622.

Webster's Unabridged Dictionary. Dorset & Baber (1983), p. 72.

Winston Universal Reference Library, The. International Press (1926–1934), p. 37.

World Book Encyclopedia, The. World Book Inc. (1988), Vol. 21, p. 299.

CHAPTER 2

Audubon Society Field Guide to North American Birds, The (Eastern Edition). Alfred A. Knopf (1977), p. 446.

Bates, Marston. *Where Winter Never Comes*. Charles Scribner's Sons (1974), p. 187.

Cohen, Daniel and M. Evans. *Intelligence: What is It?* (1974), p. 47, p. 77, p. 92, p. 94, p. 95, p. 97, p. 99, p. 108.

Donahue, Phil. *The Human Animal*. Simon & Schuster (1985), p. 324. pp. 372-375.

Douglas-Hamilton, Iain & Oria. *Among the Elephants*. Viking (1975), p. 87, p. 220.

Ehrlich, Paul and Anne. *Extinction*. Random House (1981), p. 5.

Goodall, Jane Van-Lawick. *In the Shadow of Man*. Houghton-Mifflin (1971), pp. 51-53.

Johanson, Donald C. & Maitland A. Edey. *Lucy: the Beginnings of Humankind*. Warner (1981), p. 103.

Kellog, Winthrop N. *Porpoises & Sonar*. Phoenix Science Series (1962), pp. 13–15.

Ley, Willy. *Salamanders & Other Wonders*. Viking (1951-1955), pp. 67–68, p. 75.

Lovelock, J.E. *Gaia: A New Look at Life On Earth*. Oxford University Press (1979), p. 149.

Lilly, John Cunningham. *Man & Dolphin*. Pyramid (1961), pp. 14-17.

Lilly, John Cunningham. *The Mind of the Dolphin*. Avon (1967), p. 121.

Linden, Eugene. *Apes, Men & Language*. E. P. Dutton & Co. Inc. (1974), p. 14, pp. 93–94, pp. 254–255.

Lorenz, Konrad Z. *King Solomon's Ring*. Signet (1952), p. 133.

Morris, Desmond. *The Human Zoo*. Dell (1971), pp. 25–26.

Sanderson, Ivan T. *Mammals of the World*. Doubleday & Co. Inc. (1972), pp. 216–217, p. 245.

Singer, Peter. *The Expanding Circle: Ethics & Sociobiology*. Farrar, Straus & Giroux (1981), pp. 49–52, pp. 133–134.

"Sixty Minutes." CBS (late spring, 1975). Stenuit, Robert. *The Dolphin, Cousin to Man*. Bantam (1968), pp. 59–65, pp. 72–73.

World Educator Encyclopedia The. International Book Co. (1964), Vol. 5.

Yves-Cousteau, Jacques. *The Whale*. Doubleday (1972), pp. 173–174.

CHAPTER 3

Benét, William Rose. *Reader's Encyclopedia*. Thomas Y. Crowell (1965), p. 396.

Book of Popular Science, the. The Grolier Society (1924-1931), Vol. 3, p. 797, Vol. 11, p. 3594.

Brown, Anthony. *Great Ideas in Communications*. David White Co. (1968), p. 19.

Campbell, Joseph and Bill Moyers. *The Power of Myth*. Doubleday (1988), pp. 78–79.

Comfort, William Wistar. *Quakers in the Modern World*. MacMillan Co. (1949), p. 191.

Douglas-Hamilton, Iain & Oria. *Among the Elephants*. Viking (1975), pp. 215–218.

Encyclopedia Britannica. William Benton, Publisher (1965), Vol. 5, p. 451.

Foster, G. Allen. *Communication: From Primitive Tom-Toms to Telstar*. Criterion (1965), p. 19.

Hall, Leonard. *Earth's Song*. University of Missouri Press (1981), p. 233.

Ley, Willy. *Dragons in Amber*. (1949-1951), p. 190.

Lilly, John C., M.D. *Man & Dolphin*. Pyramid (1961), pp. 59-60.

———. *The Mind of the Dolphin*. Avon (1967), p. 98.

Linden, Eugene. *Apes, Men, & Language*. E.P. Dutton (1974), p. 16, pp. 18-19, pp. 32-33.

Lorenz, Konrad Z. *King Solomon's Ring*. Thomas Y. Crowell (1952), pp. 44–45, pp. 94–95.

Morgan, Elaine. *The Descent of Woman*. Bantam (1973), pp. 129–130.

Morris, Desmond. *The Human Zoo*. Dell (1971), pp. 106–110.

Mowat, Farley. *Never Cry Wolf*. Dell (1963), pp. 60–63.

National Wildlife. "Loulis, The Talking Chimp." Mike Toner (Feb.-Mar., 1986), p. 24.

Safire William. *What's the Good Word?* Avon (1982), pp. 246–254.

Sandars, N.K. *Epic of Gilgamesh*. Penguin (1960-1972), p. 48.

259

Sanderson, Ivan T. *Living Mammals of the World*. Doubleday (1972), p. 217, p. 248, p. 257.

Shreeve, James. *Nature: The Other Earthlings*. Macmillan (1987), p. 170.

Stenuit, Robert. *The Dolphin, Cousin to Man*. Bantam (1968), pp. 56–57, p. 68, pp. 124–125.

Time magazine. "Science" column (5-2-83).

Webster's Unabridged. Dorset & Baber (1983), p. 367.

Who's Who in America (44th Ed.). MacMillian & Co. Inc. (1986), Vol. 2, p. 2532.

CHAPTER 4

Audubon Society Field Guide to North American Birds (Eastern Region). Knopf (1977), p. 410, p. 578, p. 581.

Audubon Society Field Guide to North American Mammals. Knopf (1980), pp. 346–347, pp. 542–544, pp. 557–559.

Brzowsky, Sarah. "Don't Dry Those Tears." *Parade* (11-18-84).

Evans, Howard Ensign. *Life on a Little-Known Planet*. E.P. Dutton (1978), p. 119, pp. 125–127, p. 134.

Harrowsmith. (Sept.-Oct., 1988), p. 116.

Hickman, Cleveland P., Ph.D. *Integrated Principles of Zoology*. C.V. Mosby Co. (1966), pp. 368–369.

Hopf, Alice L. *Nature's Pretenders*. G.P. Putnam & Sons (1979), p. 85.

Ley, Willy. *Salamanders & Other Wonders*. Viking (1955), p. 32.

Lorenz, Konrad Z. *Man Meets Dog*. Penguin (1953), pp. 63–65, pp. 106–110, pp. 174–176.

Milne & Milne. *The Secret Life of Animals*. E.P. Dutton, p. 18, p. 28.

Morgan, Elaine. *Descent of Woman*. Bantam (1973), pp. 24–26, pp. 39–40, pp. 43–47, pp. 129–130, pp. 147–148.

Naughton, Bill. *Alfie Darling*. MacGibbon & Kee (1966), p. 170.

Sanderson, Ivan T. *Living Mammals of the World*. Doubleday (1972), p. 63, p. 161.

Shreeve, James. *Nature: The Other Earthlings*. MacMillan (1987), pp. 36-37.

Sinclair, Sandra. *How Animals See*. Facts On File (1985), p. xii (intro.), p. 9, p. 21, pp. 68-69, p. 84, p. 88, pp. 103-104, pp. 112, p. 113, p. 115, pp. 118-119, p. 125 (photo caption), p. 127, p. 132.

Taber, C.W. *Taber's Cyclopedic Medical Dictionary*. F.A. Davis Co. (1950), p. B-50.

Southern Outdoors: "A Fish-Eye Look At Colors." (April, 1987).

Walls, G.L. *The Vertebrate Eye and Its Adaptive Radiation*. Cranbrook Press (1942), pp. 52-53, p. 463.

CHAPTER 5

Asimov, Isaac. *A Choice of Catastrophes.* Fawcett-Columbine (1979), p. 241–242, p. 257, pp. 330–337.

Associated Press. "Tilis." (5-8-88).

Audubon Guide to North American Mammals. Knopf (1980), pp. 356-357, p. 603.

Beck, Sc.D., Alan and Aaron Katcher, M.D., *Between Pets & People.* G.P. Putnam's Sons (1983), pp. 73-77.

Begley, Sharon. "Nature's Baby Killers." *Newsweek.* (9-6-82).

Book of Popular Science, The. The Grolier Society (1924-1931), Vol. 3, p. 1032, Vol. 4, p. 1207.

Dröscher, Vitus B. *They Love and Kill.* Dutton (1976), p. 23, p. 34, p. 53, p. 140, pp. 258–264.

Frazer, James G. *The Golden Bough.* Avenel (Crown Publishing) (1981), pp. 34-35, pp. 92-93.

Goldstein, Philip. *Genetics is Easy.* Lantern Press (1947-1967), p. 174, p. 252.

Johanson, Donald and Maitland Edey. *Lucy: The Beginnings of Humankind.* Warner (1981), p. 143.

Laycock, George. *The Alien Animals.* Natural History Press (1966), pp. 174–179.

Leahy, Christopher. *The Birdwatcher's Companion.* Hill & Wang (1982), p. 137, p. 154, p. 340, p. 598, pp. 608–609.

Ley, Willy. *The Lungfish, The Dodo & The Unicorn.* Viking (1941-1948), pp. 202-203.

Lorenz, Konrad Z. *King Solomon's Ring.* Thomas Y. Crowell (1952), pp. 128-141.

McLoughlin, John C. *The Canine Clan.* Viking (1983), p. 11, p. 91, p. 99.

McNeill, William H. *Plagues & Peoples.* Anchor Press (1976), p. 45, p. 82.

Morgan, Elaine. *The Descent of Woman.* Bantam (1973), p. 178.

Morris, Desmond. *The Naked Ape.* Dell (1967), p. 77.

———. *The Human Zoo.* Dell (1971), p. 72, pp. 123-124, p. 142.

Mowat, Farley. *Never Cry Wolf.* Dell (1963), p. 130.

Parade. "Homosexual Sheep?" Intelligence Report (3-8-92).

Rood, Ronald. *Animals Nobody Loves.* Bantam (1972), p. 67.

Routh, Guy. *The Origin of Economic Ideas.* Vintage (1975), pp. 107–114.

Salten, Felix. *Bambi.* Simon & Schuster (1929).

Sanderson, Ivan T. *Living Mammals of the World.* Doubleday (1972), p. 99, p. 104, p. 123, p. 222.

Shepardson, Mary. "Old Beck, the Fertile Mule." California Horse Review (Dec. 1981), Texas A & M, p. 44.

Singer, Peter. *The Expanding Circle*. Farrar, Straus & Giroux (1981), p. 58, p. 69.

Tsu, T.C. and Kurt Benirschke. *An Atlas of Mammalian Chromosomes*. Springer-Verlag (1967), Vol. I, folio 20.

Webster, Gary. *Codfish, Cats & Civilization*. Doubleday (1959), p. 59.

CHAPTER 6

Audubon Society Field Guide to North American Mammals, the. Knopf (1980), pp. 639–641.

Beck, Sc.D., Alan and Aaron Katcher, M.D. *Between Pets & People*. Putnam (1983), pp. 187–192, pp. 190–191.

Ehrlich, Paul & Anne. *Extinction*. Random House (1981), p. 69, pp. 88–89, p. 98, pp. 142–143.

Encyclopedia Americana. Grolier, Inc. (1981-1983), Vol. I, pp. 886–887.

Hopf, Alice L. *Nature's Pretenders*. (1979), pp. 76–82.

Johanson, Donald and Maitland Edey. *Lucy: The Beginnings of Humankind*. Warner (1981), p. 20, p. 24.

Laycock, George. *The Alien Animals*. Natural History Press (1966), pp. 61–68.

Ley, Willy. *Dragons in Amber*. Viking (1951), pp. 260–266.

Lorenz Konrad Z. *King Solomon's Ring*. Thomas Y. Crowell (1952), pp. 132–133.

————. *Man Meets Dog*. Penguin (1953), pp. 17–19.

MacDonald, John D. *The Lonely Silver Rain*. Knopf (1985), p. 46.

McLoughlin, John C. *The Canine Clan*. Viking (1983), p. 72, p. 73, p. 84, p. 87, p. 88, p. 89, p. 93, p. 134.

Morris, Desmond. *The Naked Ape*. Dell (1967), pp. 178–179, p. 181.

Sanderson, Ivan T. *Living Mammals of the World*. Doubleday (1972), pp. 163–165, p. 197, p. 203, p. 208, pp. 243–245, p. 292.

Tannahill, Reay. *Food in History*. Stein & Day (1973), pp. 38-62.

Webster, Gary. *Codfish, Cats & Civilization*. Doubleday (1959), pp. 14–18.

CHAPTER 7

Audubon Society Field Guide to North American Mammals, The. Knopf (1980), p. 182, p. 188, p. 191, p. 192, p. 195, pp. 384–386, pp. 457–459, pp. 591–593, pp. 611–612.

Goodall, Jane Van-Lawick. *In the Shadow of Man*. Houghton-Mifflin (1971), pp. 50–51, p. 194.

Hall, Leonard. *Earth's Song*. University of Missouri Press (1981), p. 244.

Leahy, Christopher. *The Birdwatcher's Companion*. Hill & Wang (1982), pp. 494–495.

Ley, Willy. *Salamanders & Other Wonders*. Viking (1951-1955), p. 235, pp. 236–237, p. 242, pp. 245–246.

Linden, Eugene. *Apes, Men, & Language*. E.P. Dutton (1962-1970), p. 225, p. 226, pp. 230–237.

Lorenz, Konrad Z. *King Solomon's Ring*. Thomas Y. Crowell (1952), pp. 197–203.

MacDonald, John D. *The House Guests*. Fawcett (1965), p. 51.

McLoughlin, John C. *The Canine Clan*. Viking (1983), pp. 69–70.

Mowat, Farley. *Never Cry Wolf*. Dell (1963), pp. 13–14.

Sanderson, Ivan T. *Living Mammals of the World*. Doubleday (1972), pp. 117–118, pp. 196–197.

White, E.B. *The Points of My Compass*. Harper & Row (1954-1962), pp. 63–65, p. 73.

World Educator Encyclopedia. International Book Co. (1964), Vol. 8.

Chapter 8

Bates, Marston. *Where Winter Never Comes*. Charles Scribner's Sons (1952), pp. 154–158.

Blum, David. "Polygraphs..." *Tulsa World News* (4-17-88).

Encyclopedia Americana. Grolier Inc. (1988), Vol. 25, p. 359, Vol. 29, p. 557.

Encyclopedia Britannica. Grolier (1985), Vol. 8, p. 505.

Donahue, Phil. *The Human Animal*. Simon & Schuster (1985), p. 248.

Dorset & Baber. *Webster's Unabridged*. Simon & Schuster (1955-1983), p. 867.

Douglas-Hamilton, Iain & Oria. *Among the Elephants*. Viking Press (1975), pp. 240–244.

Eisely, Loren. *The Star Thrower*. Quadrangle (1978), p. 112, pp. 251–257.

Goodall, Jane Van-Lawick. *In the Shadow of Man*. Houghton-Mifflin (1971), pp. 92–95, p. 194.

Hall, Leonard. *Earth's Song*. University of Missouri (1981), p. 213.

Johanson, Donald C. and Maitland Edey. *Lucy: The Beginnings of Humankind*. Warner (1981), pp. 29–30, pp. 40–45, pp. 63–69.

Ley, Willy. *Salamanders & Other Wonders*. Viking (1951-1955), p. 58.

Lorenz Konrad Z. *King Solomon's Ring*. Signet (1952), pp. 197–198.

Morris, Desmond. *The Human Zoo*. Dell (1971), p. 25, p. 103.

"Nova." Transcript # 1502 "Children of Eve." WGBH Educational Foundation (1986), pp. 12–13.

Raeburn, Michael, ed. *Architecture of the Western World*. Crescent (1980), p. 64, p. 67.

Serpell, James. *In the Company of Animals.* Basil Blackwell Ltd. (1986), pp. 146–147.

Singer, Peter. *The Expanding Circle.* Farrar, Straus & Giroux (1981), p. 27, p. 45, pp. 133–134, pp. 155–156.

CHAPTER 9

Asimov, Isaac. *A Choice of Catastrophes.* Fawcett-Columbine (1979), p. 223, pp. 246–247.

Audubon Society Field Guide to North American Mammals. Alfred A. Knopf (1980), p. 649.

Alternatives to Animal Use in Research, Testing, and Education. Congress of the United States, Office of Technology Assessment (1986), pp. 90–91.

Book of Popular Science, The. Grolier Society (1924-1931), Vol. 10, p. 3202, & Vol. 15, pp. 5271–5272.

Dorset & Baber. *Webster's Unabridged.* Simon & Schuster (1955-1983), p. 2059.

Ehrlich, Paul & Anne. *Extinction.* Random House (1981), pp. 78–79, p. 120.

"Florida officials shoot 10-foot alligator after a 4-year-old girl dragged into lake." (dateline, Englewood, FL.) (6-6-88).

Encyclopedia Americana. Grolier Inc. (1927-1983), Vol. 31, p. 234.

Encyclopedia of Associations (17th Ed.). Gale Research Co. (1983), Vol. 1, p. 769.

Encyclopedia of Science & Technology. McGraw-Hill (1960-1982), Vol. 10, p. 603.

Evans, Howard Ensign. *Life on a Little-Known Planet.* E.P. Dutton (1978), pp. 195–220.

Ley, Willy. *The Lungfish, The Dodo, & The Uuicorn.* Viking (1948), p. 209, p. 293.

McCabe, Katie. "Who Will Live, Who Will Die?" *Washingtonian* (August, 1986).

McCoy, J.J. *Wild Enemies.* Hawthorne (1974), p. 196.

McLoughlin, John C. *The Canine Clan.* Viking (1983), p. 132.

Midgley, Mary. *Beast & Man.* Cornell University Press (1978), p. 217.

Morris, Desmond. *The Human Zoo.* Dell (1971), pp. 64–65.

"Non-Sport Enthusiasts Seek Funds." (dateline, Washington.) Associated Press (12-14-86).

"Nova." Aids (November, 1986). PBS (first aired 1981).

Packard, Vance. *The Human Side of Animals.* Pocket (1951), p. 89.

Rowan, Andrew N. *Of Mice, Models, and Men.* State University of New York Press, (Albany) (1984), p. 119.

Ryder, Richard D. *Victims of Science.* Davis-Poynter Ltd. (1975-1983), p. 117, p. 118.

Sanderson, Ivan T. *Living Mammals of the World* (8th Ed.). Doubleday (1972), pp. 103–104, p. 125, pp. 200–202, p. 223, p. 251.

Serpell, James. *In the Company of Animals*. Basil Blackwood Ltd. (1986), p. 85, pp. 122–124.

Skorupa, Joe. *Popular Mechanics*. "Outdoors." (March, 1988), p. 32.

Walters, Robert (syndicated columnist). (11-23-87).

CHAPTER 10

Begley, Sharon. "Why is a Lefty So Different?" *Newsweek* (8-30-82).

Bronowski, J. *The Ascent of Man*. Little, Brown, & Co. (1973), p. 19.

Contemporary Authors. Gale Research Co. (1975), Vol. 53-56, p. 68.

Eisely, Loren. *The Star Thrower*. Quadrangle (1978), p. 84, p. 208, p. 212.

"Elephant: Lord of the Jungle." "Nature," (PBS) (aired 3-20-88).

Goodall, Jane Van-Lawick. *In the Shadow of Man*. Houghton-Mifflin (1971), pp. 1–48.

————. "A Plea for the Chimps." *New York Times Magazine* (5-17-87).

Ley, Willy. *The Lungfish, The Dodo, & The Unicorn*. Viking (1941-1948), p. 38, p. 301.

————. *Salamanders & Other Wonders*. Viking (1951-1955), p. 63, p. 68, p. 100.

Lilly, John C. *Man and Dolphin*. Pyramid (1961), pp. 151–156.

Lorenz, Konrad Z. *King Solomon's Ring*. Signet (1952), pp. 195–197, pp. 203–204.

McLoughlin, John C. *The Canine Clan*. Penguin (1983), p. 37, p. 40, p. 111.

Milne & Milne. *The Secret Life of Animals*. Dutton (no date listed), p. 34.

Morris, Desmond. *The Naked Ape*. Dell (1967), pp. 157–159.

————. *The Human Zoo*. Dell (1969), p. 64.

Morgan, Elaine. *The Descent of Woman*. Bantam (1973), pp. 125–126.

Sanderson, Ivan T. *Living Mammals of the World* (8th Ed.). Doubleday (1972), pp. 102–103, p. 204, p. 292.

"Trapper mauled by injured alligator." Associated Press (7-31-88).

CHAPTER 11

Beck, Sc.D., Alan & Aaron Katcher, M.D. *Between Pets & People*. G.P. Putnam's Sons (1983), pp. 27-32, p. 90, pp. 121-123, p. 264, p. 289.

Benét, William R. *Benét's Reader's Encyclopedia* (2nd Ed.). Thomas Y. Crowell (1965), p. 175.

Encyclopedia Americana. Grolier Inc. (1927-1983), Vol. 31, p. 734.

Encyclopedia of Associations (17th Ed.). Gale Research Co. (1983), Vol. I, p. 769.

Gallico, Paul and Mathemata Anstalt. *Honorable Cat.* Crown (1972), pp. 12-14, p. 72, p. 161.

Lilly M.D., John C. *Man and Dolphin.* Pyramid (1961), pp. 64-66.

Lorenz, Konrad Z. *King Solomon's Ring.* Signet (1952), p. 23, p. 96.

—————. *Man Meets Dog.* Penguin (1953), p. 67, pp. 75-76, pp. 107-108, p. 133, pp. 139-142, p. 166, pp. 179-180, pp. 184-187.

MacDonald, John D. *The House Guests.* Fawcett (1965), pp. 61-62, pp. 119-120, pp. 172-173.

McLoughlin, John C. *The Canine Clan.* Viking (1983), p. 83, p. 91, p. 96, p. 98, p. 100, p. 102, p. 106, p. 117, p. 119, p. 130.

Midgley, Mary. *Beast & Man.* Cornell University Press (1978), pp. 232-233.

Morris, Desmond. *The Naked Ape.* Dell (1967), p. 179, pp. 183-195.

—————. *The Human Zoo.* Dell (1969), pp. 64-65.

Serpell, James. *In the Company of Animals.* Basil Blackwood Ltd. (1986), p. 11, p. 46, p. 81, p. 96, p. 112, p. 113.

Timberlake, Cotton. "Pet Industry a Growing Business." Associated Press (6-85).

Chapter 12

Audubon Society Field Guide to North American Birds (Eastern Region). Alfred A. Knopf, Inc. (1977), p. 384.

Audubon Society Field Guide to North American Mammals. Alfred A. Knopf, Inc. (1980), p. 629, p. 666.

Ehrlich, Paul & Anne. *Extinction.* Random House (1981), pp. 11-12, pp. 20-22, p. 27, p. 58, pp. 98-100, pp. 236-237.

Encyclopedia Americana. Grolier Inc. (1985), Vol. 28, p. 743.

Hickman, Ph.D., Cleveland P. and C. V. Mosby *Integrated Principles of Zoology* (3rd Ed.). (1966), p. 13, p. 249.

Ley Willy. *The Lungfish, The Dodo, & The Unicorn.* Viking (1941-1948), p. 182, pp. 252-254.

Lovelock, J.E. *Gaia: A New Look at Life on Earth.* Oxford University Press (1979), p. vii, p. x, p. 3, pp. 14-32, p. 42, p. 63, p. 145, p. 152.

McClung, Robert M. *Lost Wild America.* Morrow (1969), pp. 108-111.

McCoy, J.J. *Wild Enemies.* Hawthorne (1974), p. 196.

Midgley, Mary. *Beast & Man.* Cornell University Press (1978), pp. 16-17.

World Book Encyclopedia. World Book Inc. (1985), Vol. 21, p. 249.

GENERAL RESEARCH

Baky, John S. *Humans & Animals.* Wilson (1980).

Beck, Sc.D., Alan and Aaron Katcher, M.D. *Between Pets & People.* Putnam (1983).

Fox, M.W. *Concepts in Ethology: Animal & Human Behavior.* University of Minnesota Press (1974).

Hicks, Bernice E. *All the World is Kin.* Naturegraph (1982).

Linden, Eugene. *Apes, Men, & Language.* E.P. Dutton & Co. (1974).

Lorenz, Konrad Z. *Studies in Animal & Human Behavior.* Harvard University Press (1971).

————. *King Solomon's Ring.* Harper & Row (1979), (originally published by Thomas Y. Crowell Co. Inc. (1952)).

Louis, David. *Our Animal Brothers.* Hiawatha Bondurant (1986).

Magell, Charles R. *A Bibliography on Animal Rights & Related Matters.* University Press of America (1981).

Midgley, Mary. *Beast & Man.* Cornell University Press (1978).

Morgan, Elaine. *The Descent of Woman.* Bantam (1973).

Morris, Desmond. *The Naked Ape.* Dell (1967 & 1980).

————. *The Human Zoo.* Dell (1969).

Packard, Vance. *The Human Side of Animals.* Pocket (1951).

Rowan, Andrew N. *Animals & People Sharing the World.* University Press of New England (1988).

Savesky, Kathleen and Vanessa Malcarne, eds. *People & Animals: A Human Education Curriculum Guide* (4 volumes). NAAHE (1981).

Serpell, James. *In the Company of Animals: A Study of Human-Animal Relationships.* Basil Blackwell Ltd. (1986).

Shreeve, James. *Nature: The Other Earthlings.* MacMillan (1987).

ACKNOWLEDGMENTS

The author gratefully acknowledges the assistance of the following:

- K. Benirschke, M.D.: (Prof. of Zoology, San Diego Zoological Society, Calif.)
- T.D. Bunch: (Utah State University, Dep. of Animal, Dairy & Veterinary Sciences, Utah)
- John Calardi: (Dep. of Chemistry, Cornell U., New York)
- Dr. Thomas Eisner: (Dep. of Neuro-Biology & Science, Cornell U., New York)
- Dr. Robert Gwin: (Veterinary Opthalmologist Consultant, Oklahoma City, OK)
- Dr. Riis: (Veterinary Opthalmologist, Cornell U. Small Animal Clinic, New York)
- Oliver A. Ryder, Ph.D.: (Geneticist, San Diego Zoological Society, CA)

Special thanks to the following individuals:

R. Sheila Francis, Scott (Cornell, 1982), M.J. Williams, Veda Jones, Bill Silvester, J.M. Wilson, Karen Myers, J.S. Thalheimer

Additional thanks to the following organizations:

- American Donkey & Mule Society—Route 5, Denton, TX 76201
- Foundation for Biomedical Research—818 Connecticut Ave. NW, Suite 303, Washington, DC 20006

- Fund for Animals—200 West 57th St., New York, NY 10019
- NAPPS (National Association for the Preservation and Perpetuation of Storytelling)—PO Box 309, Jonesboro, TN 37659
- PETA (People for the Ethical Treatment of Animals), PO Box 42516, Washington, DC 20015
- Everyone at the Rogers Hough Memorial Library, Rogers, AR, 72756
- ZPG (Zero Population Growth)—1400 Sixteenth St. NW, Suite 320, Washington, DC 20036

The author is confident that she has forgotten to mention at least a few people and wishes to express her sincere thanks to them.